The Water Buffalo:
NEW PROSPECTS FOR AN UNDERUTILIZED ANIMAL

Report of an Ad Hoc Panel of the
Advisory Committee on Technology Innovation
Board on Science and Technology for International Development
Commission on International Relations
National Research Council

Books for Business
New York - Hong Kong

The Water Buffalo:
New Prospects for an Underutilized Animal

by
National Research Council

ISBN: 0-89499-193-0

Reprinted from the 1981 edition

Books for Business
New York - Hong Kong
http://www.BusinessBooksInternational.com

Panel on Water Buffalo

HUGH POPENOE, Director, International Programs in Agriculture, University of Florida, Gainesville, Florida, USA, *Chairman*

STEVE P. BENNETT, Bennett's Animal Clinic, Port of Spain, Trinidad, West Indies

CHARAN CHANTALAKHANA, Dean, Department of Animal Science, Kasetsart University, Bangkok, Thailand

DONALD D. CHARLES, Senior Lecturer, Meat Production, Department of Animal Husbandry, University of Queensland, St. Lucia, Queensland, Australia

W. ROSS COCKRILL, Livestock Expert, Food and Agriculture Organization (retired), Rome, Italy

WYLAND CRIPE, Department of Veterinary Science, University of Florida, Gainesville, Florida, USA

TONY J. CUNHA, Dean, School of Agriculture, California Polytechnic University, Pomona, California, USA

A. JOHN DE BOER, Winrock International Livestock Research and Training Center, Morrilton, Arkansas, USA

MAARTEN DROST, Department of Veterinary Science, University of Florida, Gainesville, Florida, USA

MOHAMED ALI EL-ASHRY, Department of Animal Production, Faculty of Agriculture, Shoubra el Khiema, Cairo, Egypt

ABELARDO FERRER D., Former Director, Breeding Administration Centers, Animal Sciences Division, Ministry of Agriculture, Caracas, Venezuela

JEAN K. GARDNER, Former Chief, Logistics Division USAID/Philippines, Jackson, Mississippi, USA

JAMES F. HENTGES, Department of Animal Science, University of Florida, Gainesville, Florida, USA

JACK HOWARTH, School of Veterinary Medicine, University of California, Davis, California, USA

NELS M. KONNERUP, Livestock Expert, Agency for International Development (retired), Washington, D.C., USA

A. P. LEONARDS, Water buffalo breeder, Lake Charles, Louisiana, USA

JOHN K. LOOSLI, Department of Animal Science, University of Florida, Gainesville, Florida, USA

JOSEPH MADAMBA, Director, Asian Institute of Aquaculture, Iloilo, Philippines

CRISTO NASCIMENTO, Director, Brazilian Agriculture Research Corporation (EMBRAPA), Agriculture Research Center for Humid Tropics (CPATU), Belém, Pará, Brazil

DONALD L. PLUCKNETT, Consultative Group on International Agricultural Research, The World Bank, Washington, D.C., USA

WILLIAM R. PRITCHARD, Dean, School of Veterinary Medicine, University of California, Davis, California, USA

J. THOMAS REID, (deceased), Department of Animal Science, Cornell University, Ithaca, New York, USA

JOHN H. SCHOTTLER, Animal Production Officer, Department of Primary Industries, Lae, Papua New Guinea

DALBIR SINGH DEV, Dean, Animal Science Department, College of Agriculture, Agricultural University, Ludhiana, Punjab, India

ROBERT W. TOUCHBERRY, Dean, Department of Animal Science, University of Minnesota, St. Paul, Minnesota, USA

DONALD G. TULLOCH, Division of Wildlife Research, Commonwealth Scientific and Industrial Research Organisation, Winnellie, Australia

NOEL D. VIETMEYER, Board on Science and Technology for International Development, National Research Council, Washington, D.C., *Staff Study Director*

MARY JANE ENGQUIST, Board on Science and Technology for International Development, National Research Council, Washington, D.C., *Staff Associate*

Preface

The water buffalo is an animal resource whose potential seems to have been barely recognized or examined outside of Asia. Throughout the world there are proponents and enthusiasts for the various breeds of cattle; the water buffalo, however, is not a cow and it has been neglected. Nevertheless, this symbol of Asian life and endurance has performed notably well in recent trials in such diverse places as the United States, Australia, Papua New Guinea, Trinidad, Costa Rica, Venezuela, and Brazil. In Italy and Egypt as well as Bulgaria and other Balkan states the water buffalo has been an important part of animal husbandry for centuries. In each of these places certain herds of water buffalo appear to have equaled or surpassed the local cattle in growth, environmental tolerance, health, and the production of meat and calves.

Although these are empirical observations lacking painstaking, detailed experimentation, they do seem to indicate that the water buffalo could become an important resource in tropical, subtropical, and warm temperate zones in developing and developed countries.

If this is the case, then it is clear that many countries should begin water buffalo research. Serious attention by scientists could help dispel the misperceptions and uncertainties surrounding the animal and encourage its true qualities to emerge.

This report describes the water buffalo's attributes as perceived by several animal scientists. It is designed to present the apparent strengths of buffaloes compared with those of cattle, to introduce researchers and administrators to the animal's potential, and to identify priorities for buffalo research and testing.

The panel that produced this report met at Gainesville, Florida, in July 1979. It was composed of leading water buffalo experts (particularly those from outside Asia who have directed the beginnings of water buffalo industries in their countries) and leading American animal scientists, many of whom are also familiar with the animal.

This report complements *The Husbandry and Health of the Domestic Buffalo*, edited by W. Ross Cockrill and published in 1974 by the Food and Agriculture Organization of the United Nations. Cockrill's 933-page book is a "bible" of water buffalo knowledge and provides details of breeds, world distribution, physiology, and an extensive bibliography.

v

The present report is an introduction to the water buffalo and its potential. It is written particularly for decision makers, as well as scholars or students, in the hope that it will stimulate their interest in the animal and thereby increase the appreciation of, and funding for, buffalo research. The report includes much empirical observation, largely from the panel members. Some of these observations may, in the long run, prove not to be universally applicable. Much benchmark information needs to be obtained.

Since its creation in 1971, the Advisory Committee on Technology Innovation (ACTI) has investigated innovative ways to use current technology and resources to help developing countries. Often this has meant taking a fresh look at some neglected and unappreciated plant or animal species. The committee assembles ad hoc panels of experts (usually incorporating both skeptics and proponents) to scrutinize the topics selected. The panel reports serve to draw attention to neglected, but promising, technologies and resources. (For a list of ACTI reports, see page 115.) ACTI reports are provided free to developing countries under funding by the Agency for International Development (AID).

Program costs for this study were provided by AID's Office of Agriculture, Development Support Bureau, and staff support was provided by the Office of Science and Technology, Development Support Bureau.

The final draft of this report was edited and prepared for publication by F. R. Ruskin. Bibliographic editing was by Wendy D. White. Cover art was by Deborah Hanson.

Contents

1 Introduction

The domesticated water buffalo *Bubalus bubalis* numbers at least 130 million—one-ninth the number of cattle in the world. It is estimated that between 1961 and 1981 the world's buffalo population increased by 11 percent, keeping pace with the percentage increase in the cattle population.

Although there are some pedigreed water buffaloes, most are nondescript animals that have not been selected or bred for productivity. There are two general types—the Swamp buffalo and the River buffalo.

Swamp buffaloes are slate gray, droopy necked, and ox-like, with massive backswept horns that make them favorite subjects for postcards and wooden statuettes in the Far East. They are found from the Philippines to as far west as India. They wallow in any water or mud puddle they can find or make. Primarily employed as a work animal, the Swamp buffalo is also used for meat but almost never for milk production.

River buffaloes are found farther west, from India to Egypt and Europe. Usually black or dark gray, with tightly coiled or drooping straight horns, they prefer to wallow in clean water. River buffaloes produce much more milk than Swamp buffaloes. They are the dairy type of water buffalo. In India, River buffaloes play an important role in the rural economy as suppliers of milk and draft power. River buffaloes make up about 35 percent of India's milk animals (other than goats) but produce almost 70 percent of its milk. Buffalo butterfat is the major source of cooking oil (ghee) in some Asian countries, including India and Pakistan.

Although water buffaloes are bovine creatures that somewhat resemble cattle, they are genetically further removed from cattle than are the North American bison (improperly called buffalo) whose massive forequarters, shaggy mane, and small hindquarters are unlike those of cattle. While bison can be bred with cattle to produce hybrids,* there is no well-documented case of a mating between water buffalo and cattle that has produced progeny.†

---

*This is not, however, very successful; the male progeny (at least of the F_1 generation) are sterile.
†See, however, footnote page 12.

Parts of Asia and even Europe have depended on water buffaloes for centuries. Their crescent horns, coarse skin, wide muzzles, and low-carried heads
are represented on seals struck 5,000 years ago in the Indus Valley, suggesting
that the animal had already been domesticated in the area that is now India
and Pakistan. Although buffaloes were in use in China 4,000 years ago, they
are not mentioned in the literature or seen in the art of the ancient Egyptians,
Romans, or Greeks, to whom they were apparently unknown. It was not until
about 600 A.D. that Arabs brought the animal from Mesopotamia and began
moving it westward into the Near East (modern Syria, Israel, and Turkey).
Water buffaloes were later introduced to Europe by pilgrims and crusaders
returning from the Holy Land in the Middle Ages. In Italy buffaloes adapted
to the area of the Pontine Marshes southeast of Rome and the area south of
Naples. They also became established in Hungary, Romania, Yugoslavia,
Greece, and Bulgaria and have remained there ever since.

Villagers in medieval Egypt adopted the water buffalo, which has since become the most important domestic animal in modern Egypt. Indeed, during
the last 50 years, their buffalo population has doubled to over 2 million
head. The animals now supply Egypt with more meat—much of it in the form
of tender "veal"—than any other domestic animal. They also provide milk,
cooking oil, and cheese.

Other areas have capitalized on the water buffalo's promise only in very
recent times. For instance, small lots of the animals brought to Brazil (from
Italy, India, and elsewhere) during the last 84 years have reproduced so well
that they now total about 400,000 head and are still increasing, especially in
the lower Amazon region. Buffalo meat and milk are now sold widely in
Amazon towns and villages; the meat sells for the same price as beef. Nearby
countries have also become familiar with the water buffalo. Trinidad imported several breeds from India between 1905 and 1908, while Venezuela,
Colombia, and Guyana have been importing them in recent decades. During
the 1970s Costa Rica, Ecuador, Cayenne, Panama, Suriname, and Guyana introduced small herds. By 1979 the buffalo in Venezuela numbered more than
7,000 head.

Across the Pacific, the new nation of Papua New Guinea has found the
water buffalo well suited to its difficult environment. For 9 years the government has attempted to run cattle on the Sepik and Ramu Plains on Papua
New Guinea's north coast, where the temperatures are high and the forage of
poor quality. But the cattle remain thin and underweight. In the 1960s animal
scientists began evaluating water buffaloes already living in Papua New
Guinea and, encouraged by the results, introduced additional buffaloes
from Australia. These have performed remarkably well, producing greater
numbers of calves and much more meat than the cattle in the region. The
buffaloes appear to maintain appetite despite the heat and humidity, whereas
cattle do not. The government of Papua New Guinea has since imported

more water buffaloes and today has thriving herds totaling almost 3,500 head.

The United States has been slow to recognize the water buffalo's potential, but the first herd (50 head) ever imported for commercial farming arrived in February 1978.* The humble water buffalo, normally considered fit only for the steamy rice fields of Asia, is now proving itself on farm fields in Florida and Louisiana. As a result, interest in the animal is on the rise in U.S. university and farm circles.

From experience accumulated in Asia, Egypt, South America, Papua New Guinea, Australia, the United States, and elsewhere, animal scientists now perceive that many general impressions about the water buffalo are incorrect.

For example, it is widely believed that the water buffalo is mean and vicious. Encyclopedias reinforce this perception, and in the Western world it is the prevalent impression of the animal. The truth is, however, that unless wounded or severely stressed, the domesticated water buffalo is one of the gentlest of all farm animals. Despite an intimidating appearance, it is more like a household pet—sociable, gentle, and serene. In rural Asia the care of water buffaloes is often turned over to small boys and girls aged about four to nine. The children spend their days with their family's gentle buffalo, riding it to water, washing it down, waiting while it rolls and wallows, and then riding it to some source of forage, perhaps a grassy field or thicket. It is not uncommon to see a buffalo patiently feeding, with a young friend stretched prone on its broad gray back, asleep.†

Perhaps the notion about the viciousness of water buffaloes stems from confusing them with the mean-tempered African buffalo *Syncerus caffer*, actually a distant relative with which they will not interbreed and which is classified in a different genus.

Ferocity is the most blatant misconception concerning the water buffalo, although other fallacies are widely reported as well.

One generally held belief is that water buffaloes can be raised only near water. Actually, buffaloes love to wallow, but they grow and reproduce normally without it, although in hot climates they must have shade available.

Another belief is that the water buffalo is exclusively a tropical animal. River-type buffaloes, however, have been used to pull snow plows during Bulgarian winters. They are found in Italy (over 100,000 head), Albania, Yugoslavia, Greece, Turkey, the Georgia and Azerbaijan areas of the Soviet Union (almost 500,000 head) and other temperate-zone regions as well. They are also found in cold, mountainous areas of Pakistan, Afghanistan, and Nepal.

*Air-freighted from the wilds of Guam, the U.S. island possession on the western Pacific, by panel member Tony Leonards. Prior to that time (in 1974), four head of water buffalo were imported to the Department of Animal Science, University of Florida, for study. The only other water buffaloes in North America were a few animals in zoos.
†An apparent exception is the Egyptian male buffalo, which is highly temperamental.

	SWAMP BREEDS AND TYPES	MEDITERRANEAN TYPES	RIVER BREEDS
> 1 000 000			
50 000 - 1 000 000			
1 000 - 50 000			
< 1 000			

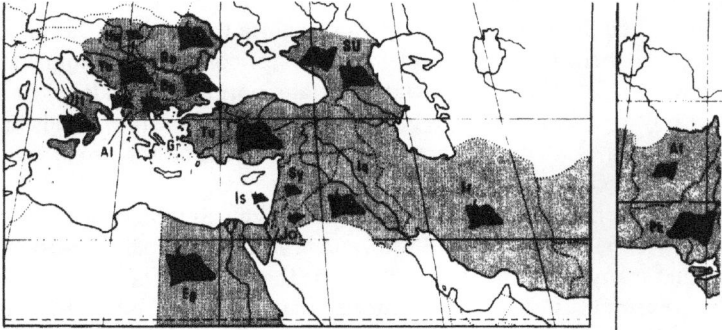

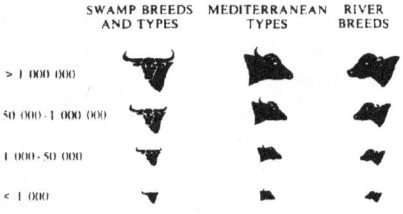

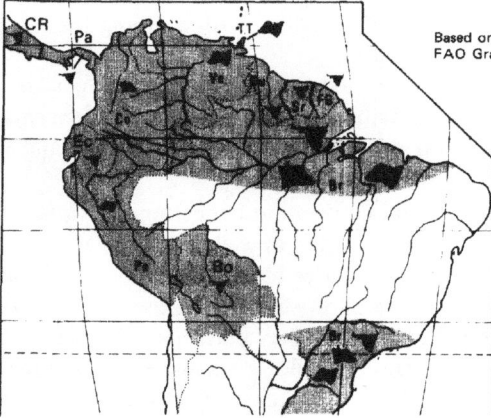

Based on Map by
FAO Graphics Section

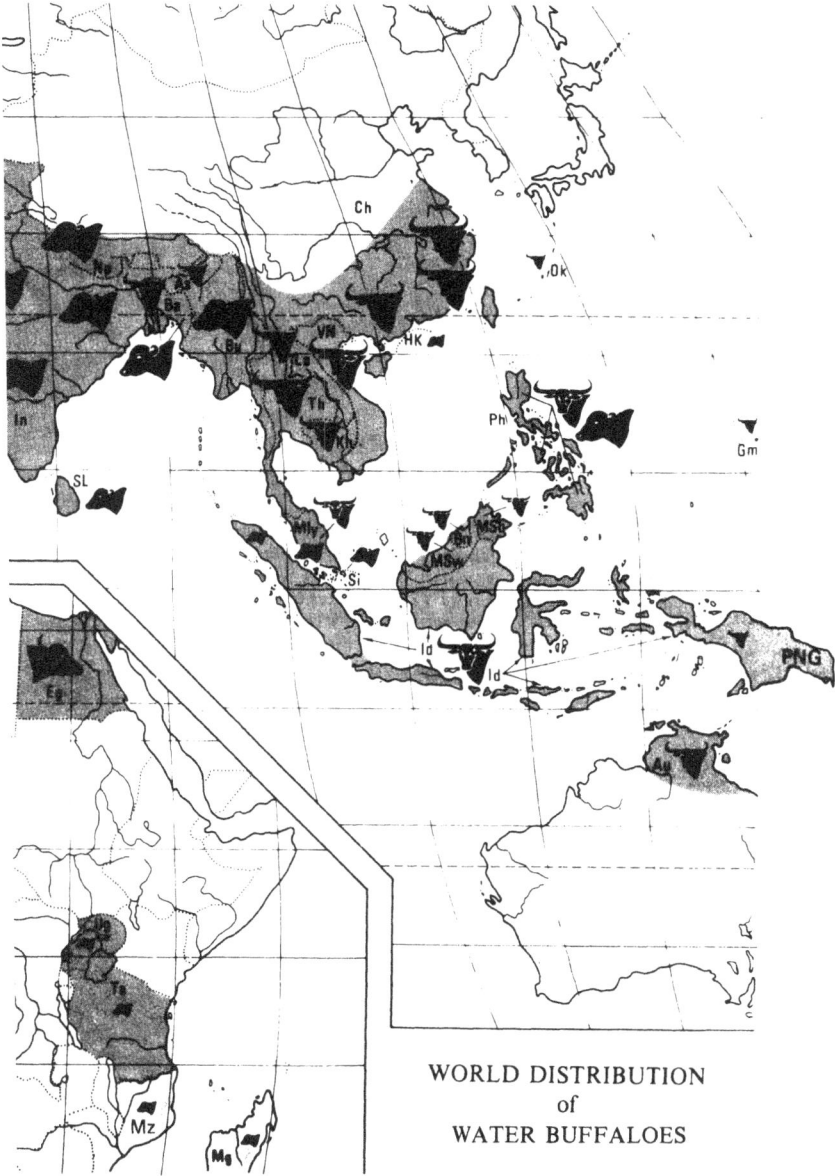

WORLD DISTRIBUTION
of
WATER BUFFALOES

Bangladesh

Swamp-type buffaloes have swept-back horns and are native to the eastern half of Asia from India to Taiwan. All are similar in general appearance. (Picture credits: Thailand, Agency for International Development; Bangladesh, B. Rumich, FAO; Vietnam, Don Hong-Oai.)

Vietnam

Murrah

Jafarabadi

Nili/Ravi

Toda

River-type buffaloes generally have curled horns and are native to the western half of Asia. Centuries ago they were introduced to Egypt and to several European countries. They have since developed distinct forms. At least 18 different "breeds" of River buffalo are known. Most are good milking animals. (Picture credits: Murrah and Italian, W. R. Cockrill; Jafarabadi, A. Santiago; Nili/Ravi, Brigadier Zafar Khan; Toda, Director of Animal Husbandry, Madras; Trinidadian, N. D. Vietmeyer.)

Italian

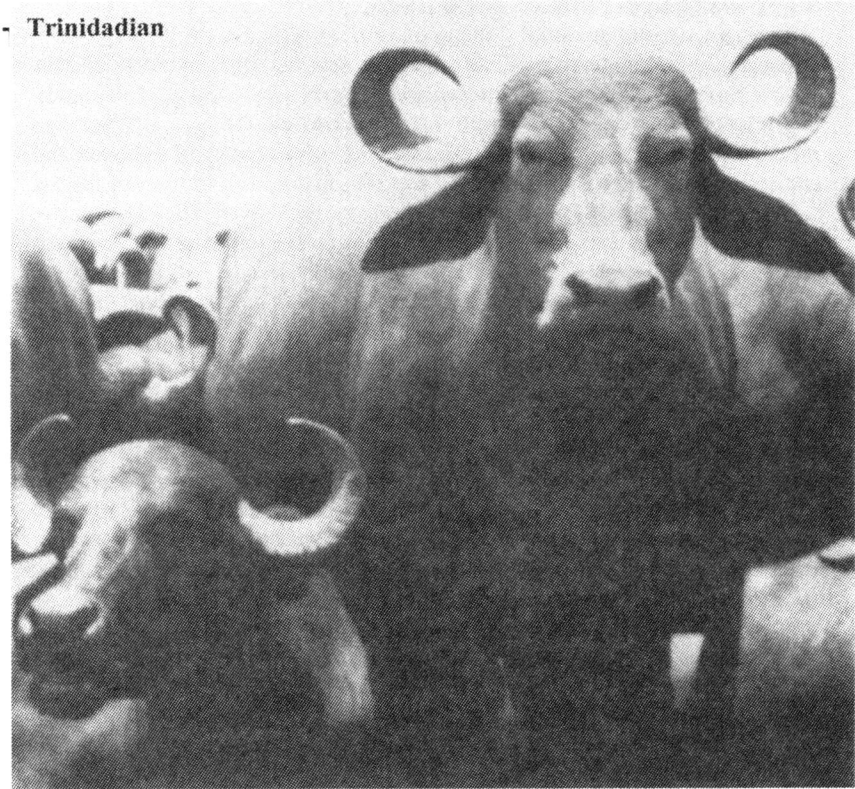

- Trinidadian

Yet another misconception is that the water buffalo is just a poor man's beast of burden. In addition to providing fine lean meat, buffaloes in fact provide rich milk. Mozzarella cheese, one of the most popular in Europe, comes from the buffaloes in Italy. Buffalo milk has a higher content of both butterfat and nonfat solids than cow's milk does. It therefore often brings higher prices than cow's milk. Throughout much of India it is in such demand that cow's milk is sometimes hard to sell.

Many of the misconceptions generally held about buffaloes are based on little data and much prejudice. For instance, it is widely believed that water buffalo meat is tough and less desirable than beef. However, when the animals are raised for meat, buffalo steaks are lean, as tender as beef, and in appearance it is difficult to distinguish the two. In taste-preference tests at the University of Queensland, buffalo steaks were preferred over those from Angus and Hereford cattle. Tests conducted in Trinidad, Venezuela, the United States, and Malaysia produced similar results.

Australia has shipped water buffalo meat to Hong Kong, the United States, Germany, and Scandinavia. Buffalo meat is now available in stores in Australia's Northern Territory, where demand exceeds supply. It sells at competitive prices and is particularly sought for barbecues and the famous Australian meat pie. In the Philippines, two-thirds of the meat consumed in homes and restaurants is actually water buffalo, a fact that many Filipinos do not realize.

Compared with cattle, water buffaloes apparently have an efficient digestive system, one which extracts nourishment from forage so coarse and poor that cattle do not thrive on it. Thin cattle are commonly seen in Asia and northern Australia, for example, but it's rare to see a protruding rib on a buffalo, even though it uses the same source of feed.

In Asia, the Middle East, and Europe, water buffaloes live on coarse vegetation on the marginal land traditionally left to the peasants. They help make human survival possible. An old Chinese woman in Taiwan once told panelist W. Ross Cockrill: "To my family the buffalo is more important than I am. When I die, they'll weep for me; but if our buffalo dies, they may starve."

A better understanding of the water buffalo could be invaluable to many developing nations. In particular, improved production of water buffalo meat offers hope for helping feed India, the second most populous nation on earth. Although India for religious reasons forbids the slaughter of cows, it has no prohibitions regarding slaughter of water buffaloes or the consumption of buffalo meat.

Most developing countries are in the tropics, and the water buffalo is inherently a tropical animal. It is comfortable in hot, humid environments. In the Amazon, for example, buffaloes are now common on the landscape and may even replace cattle completely.

Tropical countries have more serious disease problems than temperate countries do. Although susceptible to most cattle diseases, the water buffalo seems to resist ticks and often appears to be more resistant to some of the most devastating plagues that make cattle raising risky, difficult, and sometimes impossible in the tropics. Several researchers report that when water buffaloes are allowed to wallow, their mud-coated skin seems to deter insect and tick ectoparasites and they consequently require greatly reduced treatment with insecticides. Although the buffalo fly (*Siphona exigua*) affects the animals, other pests such as the warble fly and the screwworm, for example, seldom affect healthy buffaloes. Also, despite their inclination for living in swamps, rivers, and ponds, diseases of the feet such as foot rot and foot abscesses are rare.

Another benefit to developing countries is the water buffalo's legendary strength. A large part of the total farm power available in South China, Thailand, Indonesia, the Philippines, the Indochina states, India, and Pakistan comes from this "living tractor." Dependable and docile, the animals pull plows, harrows, and carts with loads that weigh several tons. In the Amazon buffalo teams pull boats laden with cargo and tourists through shallows and swamps.

The petroleum crisis has forced many farmers to reconsider animal power even in some of the technically advanced countries. Buffaloes are not only extraordinarily strong, they can also work in deep mud that may bog down a tractor. Even up to their bellies they forge on, dragging both plow and driver through the mud. Although its average walking speed is only about 3 kilometers per hour, the buffalo, unlike its mechanical competition, doesn't need gasoline or spare parts and its working life is often 20 years or more.

Breeds

As already noted, the major genetic divisions of the water buffalo are the Swamp buffalo of the eastern half of Asia, which has swept-back horns, and the River buffalo of the western half of Asia, which usually has curled horns. There is also the Mediterranean buffalo, which is of the River type but has been isolated for so long that it has developed some unique characteristics. (Records of the buffalo in Italy date back 1,000 years, during which there have been no reported imports.) Mediterranean buffaloes are stocky, high-yielding animals that combine both beef and dairy characteristics.

Although there is only one breed of Swamp buffalo, certain subgroups seem to have specific inherited characteristics. For example, the buffaloes of Thailand are noted for their large size, averaging 450–550 kg, and weights of up to 1,000 kg have been observed. Elsewhere, Swamp buffaloes range from 250 kg for some small animals in China to 300 kg in Burma and 500–600 kg in Laos.

Only in India and Pakistan are there well-defined breeds with standard qualities. There are eighteen River buffalo breeds in South Asia, which are further classified into five major groups designated as the Murrah, Gujarat, Uttar Pradesh, Central Indian, and South Indian breeds. These are the five groups and main breeds:

Group	Breeds
Murrah	Murrah, Nili/Ravi, Kundi
Gujarat	Surti, Mehsana, Jafarabadi
Uttar Pradesh	Bhadawari, Tarai
Central Indian	Nagpuri, Pandharpuri, Manda, Jerangi, Kalahandi, Sambalpur
South Indian	Toda, South Kanara

The best-known breeds are Murrah, Nili/Ravi, Jafarabadi, Surti, Mehsana, and Nagpuri. Most of the buffaloes of the Indian subcontinent belong to a nondescript group known as the Desi buffalo. There is no controlled breeding among these animals and most are quite small, yield little milk, and are variable in color.

Genetics

The Swamp buffalo has 48 chromosomes, the River buffalo, 50. The chromosomal material is, however, similar in the two types and they crossbreed to produce fertile hybrid progeny. Cattle, however, have 60 chromosomes and although mating between cattle and buffaloes is common, hybrids from the union are unlikely to occur.*

Individual buffaloes show large variation in milk yield, conformation, horn shape, color, meat production, temperament, growth rate, and other characteristics. Selection for survival under adverse conditions has occurred naturally (those that could not stand adversity died early) and farmers have probably tended to select animals of gentle temperament. But systematic genetic improvement has almost never been attempted. It seems likely that further selection could quickly improve their productivity.

Unfortunately, the large bulls that would be best for breeding purposes are often being selected as draft animals and castrated, or sent to slaughter, or (as with some feral animals in northern Australia and on the Amazon island of Marajo) shot by hunters. The result is that the buffalo's overall size in countries such as Thailand and Indonesia has been decreasing as the genes for large size and fast growth are lost.

*In 1965 a reputed hybrid was born at Askaniya Nova Zoopark in the Soviet Union (see Gray, A., 1971. *Mammalian Hybrids*, Commonwealth Agricultural Bureaux, Slough, England, p. 126). Hybrids have also been reported from China (Van Fu-Czao 1959, Gibridy buivolic i krupnogo i rogatogo skota (buffalo and cattle hybrids) *Zhivotnovodstvo*, Mosk., 21:92). Both of these reports seem doubtful because despite many attempts, no other hybrids have ever been claimed to have been produced.

Limitations

The buffalo is still largely an animal of the village, and many of its reported limitations are caused more by its environment than by the animal itself. Moreover, much of the animal's genetic potential is obscured by environmental influences. For example, for many breeds and types the genetic variations in milk yield and growth cannot be accurately determined because they are overwhelmed by the effects of inadequate nutrition and management.

Nevertheless, some inherent limitations of buffaloes can be identified. For instance, buffaloes suffer if forced to remain, even for a few hours, in direct sunlight. They have only one-tenth the density of sweat glands of cattle and their coating of hair is correspondingly sparse, providing little protection from the sun. Accordingly, buffaloes must not be driven over long distances in the heat of the day. They must be allowed time for watering and, if possible, for wallowing. Droving under a hot sun for long hours will cause heat exhaustion and possibly death; losses can be very high and can occur suddenly. Young calves are particularly affected by heat.

Buffaloes are also sensitive to extreme cold and seem less able than cattle to adapt to truly cold climates.* Sudden drops in temperature and chill winds may lead to pneumonia and death.

The water buffalo is usually found in areas where there is ready access to a wallow or shower. This is not a necessity, but when temperatures are high the availability of water is important for maintaining buffalo health and productivity. It seems clear, then, that the buffalo is not suitable for arid lands.

Increasing buffalo productivity through breed improvement is just now beginning. Throughout Asia buffalo mating is almost completely haphazard, and so the animal lacks the quality improvement through breeding that most cattle have had. Therefore, most buffaloes are of nondescript heritage and genetic potential.

On poor quality feed water buffalo grow at least as well as cattle, but under intensive conditions they probably won't grow as fast as the best breeds of cattle. In feedlots, therefore, the buffalo is likely to be less productive than improved cattle. Weight gains of about 1 kg per day have been recorded; some exceptional cattle may gain at almost twice that rate.

The buffalo has long been considered a poor breeder—slow to mature sexually, and slow to rebreed after calving. Accumulated experience now shows, however, that this is mainly a result of poor management and nutrition. Buffaloes are not sluggish breeders. Nevertheless, their gestation period is about a month longer than that of cows, buffalo estrus is difficult to detect, and many matings occur at night so that farmers are likely to encounter more problems breeding buffaloes than cattle.

---

*A rule of thumb is that buffaloes don't do well where the sun is inadequate to ripen, say, cotton, grapes, or rice. Kaleff, B., 1942. Der Hausbuffel und seine Zuchtungsbiologie im Vergleich zum Rind. *Zeitschsift Tierzucht Biologie*, 51:131–178.

Although reputed to be ferocious, water buffalo are actually docile, gentle animals. In most regions where they exist they have a close relationship with their owners, becoming almost like family pets. These intelligent animals seem to welcome the attachment. Left: Clothes and buffalo get a wash together, India. (FAO Photo) Above: Children tending water buffalo, Thailand. (UNICEF photo issued by FAO) Below: Buffalo enjoys daily rubdown, Gainesville, Florida. (N. D. Vietmeyer) Overpage: Buffalo race, Philippines. (Gina Lollobrigida)

Buffaloes are gentle creatures, but if roughly or inexpertly handled they can, through fear or pain, become completely unmanageable. Buffalo behavior sometimes differs from that of cattle. For example, most buffaloes are not trained to be driven. Instead, the herdsman must walk alongside or ahead of them; they then instinctively follow. Also, because of their innate attachment to an individual site or herd it is more difficult to move buffaloes to new locations or herds. In addition, buffaloes respect fences less than cattle do and when they have the desire to move they are harder to contain. (Electric fences, however, will stop them.)

Despite their general good health, buffaloes are probably as susceptible as cattle to most infections. However, the buffalo seems to be peculiarly sensitive to a few cattle diseases and resistant to a few others (see chapter 7). Reactions to some diseases seem to vary with region, environment, and breed, and the differences are not well understood.

Destruction of the environment is sometimes attributed to buffalo wallowing. This danger seems to have been overstated, except in cases where stocking rates were unreasonably high.* However, buffaloes rub against trees more often than cattle do, and they sometimes de-bark the trees, causing them to die.

Unfortunately, some of the best genetic stocks of water buffaloes exist in areas where certain infections and viral and other diseases sometimes occur. Thus, many countries are reluctant to permit importation of water buffaloes, despite the fact that modern quarantine procedures under conditions of maximum security can essentially eliminate the risk.

Finally, it must be emphasized that because buffalo research has been largely neglected, most results reported in this and other buffalo writings cover small numbers of animals and short periods of time. Many are merely empirical observations that have not been subjected to independent confirmation.

Selected Readings

Anonymous. 1972. Buffalo at the crossroads. *World Farming* 14(7):10-13.
Asian and Pacific Council. 1979. Priorities in buffalo research identified. *ASPAC Newsletter* No. 43.
Bowman, J. C. 1977. *Animals for Man*. Edward Arnold Publishers, Ltd., London, United Kingdom.
Buffalo Bulletin. Newsletter published by the Buffalo Research Committee of Kasetsart University, Bangkok, Thailand. (In English and Thai.)

*An ongoing study in Northern Australia of environmental degradation widely attributed to buffalo has now shown that the effects are caused by man and climatic changes and only very slightly by buffaloes. (Information supplied by D. G. Tulloch.)

Buffalo World. Newsletter published by the National Dairy Research Institute, Karnal, 132001, Haryana, India.

Chantalakhana, C. 1975. The buffaloes of Thailand – their potential, utilization and conservation. In: *The Asiatic Water Buffalo*. Proceedings of an International Symposium held at Khon Kaen, Thailand, March 31–April 6, 1975. Food and Fertilizer Technology Center, Taipei, Taiwan.

Chantalakhana, C., and Na Phuket, S. R. 1979. The role of swamp buffalo in small farm development and the need for breeding improvement in Southeast Asia. *Extension Bulletin* No. 125. Food and Fertilizer Technology Center, Taipei, Taiwan.

Cockrill, W. R. 1967. The water buffalo. *Scientific American* 217:118.

Cockrill, W. R., ed. 1974. *The Husbandry and Health of the Domestic Buffalo*. Food and Agriculture Organization of the United Nations, Rome, Italy.

Cockrill, W. R. 1975. The domestic buffalo. *Blue Book for the Veterinary Profession*, No. 25. Animal Production and Health Division, Food and Agriculture Organization of the United Nations, Rome, Italy.

Cockrill, W. R. 1976. *The Buffaloes of China*. Food and Agriculture Organization of the United Nations, Rome, Italy.

Cockrill, W. R., ed. 1977. *The Water Buffalo*. Food and Agriculture Organization of the United Nations, Rome, Italy.

Cockrill, W. R. 1977. The water buffalo: domestic animal of the future. *Bovine Practitioner* 12:92-98.

Cockrill, W. R. 1978. Domestic water buffaloes. In: *The Care and Management of Farm Animals*, edited by W. N. Scott. Bailliere Tindall, London, United Kingdom.

Cockrill, W. R. 1980. The ascendant water buffalo – key domestic animal. *World Animal Review* 33:2-13.

DeBoer, A. J. 1972. Technical and economic constraints on bovine production in three villages in Thailand. *Dissertation Abstracts International* 33(5):1935.

DeBoer, A. J. 1975. *Livestock and Poultry Industry in Selected Asian Countries. Report of Survey on Diversification of Agriculture: Livestock and Poultry Production*. Asian Productivity Organization, Tokyo, Japan.

de Guzman, M. R., Jr. 1979. An overview of recent developments in buffalo research and management in Asia. *Extension Bulletin* No. 124. Food and Fertilizer Technology Center, Taipei, Taiwan.

Documentation Center on Water Buffalo. 1978. *Abstract Bibliography on Water Buffalo, 1971-1975*. Documentation Center on Water Buffalo, University of the Philippines at Los Baños Library, College, Laguna 3720, Philippines.

Fahimuddin, M. 1975. *Domestic Water Buffalo*. Oxford and IBH Publishing Company, New Delhi, India.

Fischer, H. 1975. The water buffalo and related species as important genetic resources: their conservation, evaluation and utilization. In: *The Asiatic Water Buffalo*. Proceedings of an International Symposium held at Khon Kaen, Thailand, March 31–April 6, 1975. Food and Fertilizer Technology Center, Taipei, Taiwan.

Ford, B. D., and Tulloch, D. G. 1977. The Australian buffalo – a collection of papers. *Technical Bulletin* No. 18. Department of the Northern Territory, Animal Industry and Agriculture Branch, Australian Government Publishing Service, Canberra, Australia.

Gupta, H. C. 1977. Possibilities and realities of developing buffalo's performance, breeding and feeding. *Indian Dairyman* 29(6):337-346.

Kartha, K. P. R. 1965. Buffalo. In: *An Introduction to Animal Husbandry in the Tropics*, edited by G. Williamson and W. J. A. Payne. Longman, London, United Kingdom.

McKnight, T. L. 1971. Australia's buffalo dilemma. *Annals of the Association of American Geographers* 61(4):759-773.

Madamba, J. C., and Eusebio, A. N. 1979. Developments in the strengthening of buffalo research in Asia. *Buffalo Bulletin* 2(3):7-16.

Mahadevan, P. 1978. Water buffalo research – possible future trends. *World Animal Review* 25:2-7.

Oloufa, M. M. 1979. Buffaloes as producers of meat and milk. *Egyptian Journal of Animal Production* 19:1-10.

Pant, H. C., and Roy, A. 1972. The water buffalo and its future. In: *Improvement of Livestock Production in Warm Climates*, edited by R. E. McDowell. W. H. Freeman and Company, San Francisco, California, USA.

Philippine Council for Agriculture and Resources Research. 1978. *The Philippines Recommends for Carabao Production*, PCARR, Los Baños, Laguna 3732, Philippines.

Ranjhan, S. K., and Pathak, N. N. 1981. *Management and Feeding of Buffaloes.* Vikas Publishing House, New Delhi, India.

Robinson, D. W. 1977a. *Livestock in Indonesia.* Research Report No. 1. Centre for Animal Research and Development, Bogor, Indonesia. (In English and Indonesian.)

Robinson, D. W. 1977b. *Preliminary Observations on the Productivity of Working Buffalo in Indonesia.* Research Report No. 2. Centre for Animal Research and Development, Bogor, Indonesia. (In English and Indonesian.)

Sundaresan, D. 1979. The role of improved buffaloes in rural development. *Indian Dairyman* 31(2):73-78.

Tulloch, D. G. 1978. The water buffalo, *Bubalus bubalis*, in Australia: grouping and home range. *Australian Wildlife Research* 5:327-34.

Wahid, A. 1973. Pakistani buffaloes. *World Animal Review* 7:22-28.

2 Meat

The water buffalo offers promise as a major source of meat, and the production of buffaloes solely for meat is now expanding.

Because buffaloes have been used as draft animals for centuries, they have evolved with exceptional muscular development; some weigh 1,000 kg or more. Until recently, however, little thought was given to using them exclusively for meat production. Most buffalo meat was, and still is, derived from old animals slaughtered at the end of their productive life as work or milk animals. As a result, much of the buffalo meat sold is of poor quality. But when buffaloes are properly reared and fed, their meat is tender and palatable.

Water buffaloes are exported for slaughter from India and Pakistan to the Middle East and from Thailand and Australia to Hong Kong. Demand for meat is so great that Thailand's buffalo population has dropped from 7 million to 5.7 million head in the last 20 years, a period in which the human population has more than doubled.

Carcass Characteristics

All buffalo breeds—even the milking ones—produce heavy animals whose carcass characteristics are similar to those of cattle.

Despite heavier hide and head, the amount of useful meat (dressing percentage) from buffaloes is almost the same as in cattle. Mediterranean type buffalo and Zebu cattle steers in Brazil yielded dressing percentages of 55.5 and 56.6 percent respectively.* Swamp buffalo dressing percentages have been measured in Australia at 53 percent.†

Buffaloes are lean animals. Although a layer of subcutaneous fat covers the carcass, it is usually thinner than that on comparably fed cattle. Even animals that appear to be fat prove to be largely muscle. Australian research on Swamp buffaloes reveals that buffaloes with more than 25 percent fat are difficult to produce, whereas average choice-grade beef carcasses may contain

---

*Information supplied by C. Nascimento.
†Information supplied by D. D. Charles. Cattle in Australia average 53–56 percent. Generally buffalo have about 3 percent lower dressing percentage than cattle.

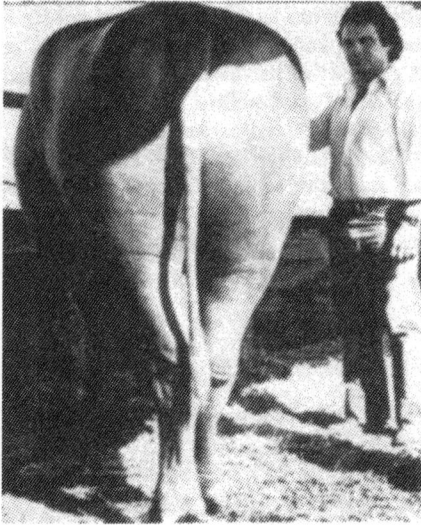

Buffaloes can be bred for good meat conformation. Above: Prize bull "Fat Boy," Trinidad, 12 months old, 320 kg. (G. Cordell) Left: Jafarabadi bull, 40 months old, 1,050 kg. Brazilian National Champion in 1980. (E. Aziz Haik)

about 35 percent fat.* This lower level of fat is sometimes seen even under feedlot conditions, although animals liberally fed concentrated rations will eventually fatten. Castrated males have a reasonably even layer of subcutaneous fat; it is often difficult to differentiate their carcasses from those of cattle steers of equivalent quality.†

In general, the buffalo carcass has rounder ribs, a higher proportion of muscle, and a lower proportion of bone and fat than beef has.

*Charles and Johnson, 1975 and Johnson and Charles, 1975. Other buffalo breeds may differ. Yugoslavian researchers have found no difference between Mediterranean buffaloes and cattle (Buska breed) in carcass leanness and physical properties (Joksimovic and Ognjanovic, 1977).
†Information supplied by D. G. Tulloch.

Buffalo hide is so thick that it can be sliced into two or three layers before tanning into leather.

Meat Quality

Buffalo meat and beef are basically similar. The muscle pH (5.4), shrinkage on chilling (2 percent), moisture (76.6 percent), protein (19 percent), and ash (1 percent) are all about the same in buffalo meat and beef. Buffalo fat, however, is always white and buffalo meat is darker in color than beef because of more pigmentation or less intramuscular fat (2-3 percent "marbling," compared with the 3-4 percent in beef).

Eating Quality

Taste-panel tests and tenderness measurements conducted by research teams in a number of countries have shown that the meat of the water buffalo is as acceptable as that of cattle. Buffalo steaks have rated higher than beefsteaks in some taste tests in Australia, Malaysia, Venezuela, and Trinidad.

In taste-panel studies in Trinidad, cooked joints from three carcasses—a Trinidad buffalo, a crossbred steer (Jamaica–Red/Sahiwal), and an imported carcass of a top-grade European beef steer—were served. The 28 diners all had experience in beef production, butchery, or catering and were not told the sources of the various joints. All the carcasses were held in cold storage for one week before cooking. The buffalo meat was rated highest by 14 judges; 7 chose the European beef; 5 thought the crossbred beef the best; and 2 said that the buffalo and crossbred were equal to or better than the European beef. The buffalo meat received most points for color (both meat and fat), taste, and general acceptability. There was little difference noted in texture.*

Buffalo veal is considered a delicacy. Calves are usually slaughtered for veal between 3 and 4 weeks of age; dressed weight is 59-66 percent of live weight.

There is some evidence that buffaloes may retain meat tenderness to a more advanced age than cattle because the connective tissue hardens at a later age or because the diameter of muscle fibers in the buffalo increases more slowly than in cattle.† In one test the tenderness (measured by shearing force) of muscle samples from carcasses of buffalo steers 16-30 months old was the same as that from feedlot Angus, Hereford, and Friesian steers 12-18 months old.‡ This gives farmers more flexibility in meeting fluctuating markets while still providing tender meat.

*Information supplied by P. N. Wilson.
†Joksimovic, 1979.
‡Charles and Johnson, 1972.

Selected Readings

Anonymous. 1976. *Livestock Production in Asian Context of Agricultural Diversification.* Asian Productivity Organization, Tokyo, Japan.

Arganosa, F. C., Sanchez, P. C., Ibarra, P. I., Gerpacio, A. L., Castillo, L. S., and Arganosa, V. G. 1973. Evaluation of carabeef as a potential substitute for beef. *Philippine Journal of Nutrition* 26(2).

Arganosa, F. C., Arganosa, V. G., and Ibarra, P. I. 1975. Carcass evaluation and utilization of carabeef. In: *The Asiatic Water Buffalo.* Proceedings of an International Symposium held at Khon Kaen, Thailand, March 31–April 6, 1975. Food and Fertilizer Technology Center, Taipei, Taiwan.

Bennett, S. P. 1973. The "buffalypso"—an evaluation of a beef type of water buffalo in Trinidad, West Indies. Paper presented at the Third World Conference on Animal Production, Melbourne, Australia.

Borghese, A., Gigli, S., Romita, A., Di Giacomo, A., and Mormile, M. 1978. Fatty acid composition of fat in water buffalo calves and bovine calves slaughtered at 20, 28, and 36 weeks of age. In: *Patterns of Growth and Development in Cattle: A Seminar in the EEC Programme of Coordination of Research on Beef Production, held at Ghent, October 11-13, 1977.* Martinus Nijhoff, The Hague, Netherlands.

Charles, D. D., and Johnson, E. R. 1972. Carcass composition of the water buffalo (*Bubalus bubalis*). *Australian Journal of Agriculture Research* 23:905–911.

Charles, D. D., and Johnson, E. R. 1975. Live weight gains and carcass composition of buffalo (*Bubalus bubalis*) steers on four feeding regimes. *Australian Journal of Agriculture Research* 26:407–413.

Cockrill, W. R. 1975. Alternative livestock: with particular reference to the water buffalo (*Bubalus bubalis*). In: *Meat. Proceedings of the 21st Easter School in Agricultural Science, University of Nottingham,* edited by D. J. A. Cole and R. A. Lawrie. Butterworth, London, England.

El-Ashry, M. A., Mogawer, H. H., and Kishin, S. S. 1972. Comparative study of meat production from cattle and buffalo male calves. *Egyptian Journal of Animal Production* 12:99–107.

El-Koussy, H. A., Afifi, Y. A., Dessouki, T. A., and El-Ashry, M. A. 1977. Some chemical and physical changes of buffalo meat after slaughter. *Agriculture Research Review* 55:1–7.

Johnson, E. R., and Charles, D. D. 1975. Comparison of live weight gain and changes in carcass composition between buffalo (*Bubalus bubalis*) and *Bos taurus* steers. *Australian Journal of Agriculture Research* 26:415–422.

Joksimovic, J. 1979. Physical, chemical and structural characteristics of buffalo meat. *Arhiv za Poljoprivredne Nauke* 22:110.

Joksimovic, J., and Ognjanovic, A. 1977. A comparison of carcass yield, carcass composition, and quality characteristics of buffalo meat and beef. *Meat Science* 1:105–110.

Mai, S. C., and Wu, T. H. 1974. TSC's intensive feed-lot system for cow-calf and beef production program. *Taiwan Sugar* 21(6):198–211.

Matassino, D., Romita, A., Cosentino, E., Girolami, A., and Cloatruglio, P. 1978. Myorheological, chemical, and colour characteristics of meat in water buffalo and bovine calves slaughtered at 20, 28 and 36 weeks. In: *Patterns of Growth and Development in Cattle: A Seminar in the EEC Programme of Coordination of Research on Beef Production, held at Ghent, October 11-13, 1977.* Martinus Nijhoff, The Hague, Netherlands.

Mogawer, H. H., El-Ashry, M. A., and Mahmoud, S. A. 1976. Comparative study of meat production from cattle and buffalo male calves. II. Effect of different roughage concentrate ratios in ration on carcass traits. *Journal of Agriculture Research (Tanta University)* 2:6–12.

Ognjanovic, A. 1974. Meat and meat production. In: *The Husbandry and Health of the Domestic Buffalo,* edited by W. R. Cockrill. Food and Agriculture Organization of the United Nations, Rome, Italy.

Rastogi, R., Youssef, F. G., and Gonzalez, F. D. 1978. Beef type water buffalo of Trinidad-Beefalypso. *World Review of Animal Production* 14(2):49-56.

Romita, A., Borghese, A., Gigli, S., and Di Giacomo, A. 1978. Growth rate and carcass composition of water buffalo calves and bovine calves slaughtered at 20, 28 and 36 weeks. In: *Patterns of Growth and Development in Cattle: A Seminar in the EEC Programme of Coordination of Research on Beef Production, Held at Ghent, October 11-13, 1977.* Martinus Nijhoff, The Hague, Netherlands.

Wilson, P. N. 1961. Palatability of water buffalo meat. *Journal of the Agricultural Society of Trinidad* 61:457, 459-460.

3 Milk

More than 5 percent of the world's milk comes from water buffaloes. Buffalo milk is used in much the same way as cow's milk. It is high in fat and total solids, which gives it a rich flavor. Many people prefer it to cow's milk and are willing to pay more for it. In Egypt, for example, the severe mortality rate among buffalo calves is due in part to the sale of buffalo milk, which is in high demand, thus depriving calves of proper nourishment. This also occurs in India, where in the Bombay area alone an estimated 10,000 newborn calves starve to death each year through lack of milk. The demand for buffalo milk in India (about 60 percent of the milk consumed; over 80 percent in some states) is reflected in the prices paid for a liter of milk: about 130 paisa for cow's milk compared with about 200 paisa for buffalo milk.

Twelve of the 18 major breeds of water buffalo are kept primarily for milk production (although males may be used for traction and all animals are eventually used for meat). The main milk breeds of India and Pakistan are the Murrah, Nili/Ravi, Surti, Mehsana, Nagpuri, and Jafarabadi. The buffaloes of Egypt, Eastern Europe (Bulgaria, Romania, Yugoslavia, and the USSR), and Italy are used for milk production and there are also herds used principally for this purpose in Iran, Iraq, and Turkey.

Composition

Buffalo milk contains less water, more total solids, more fat, slightly more lactose, and more protein than cow's milk. It seems thicker than cow's milk because it generally contains more than 16 percent total solids compared with 12-14 percent for cow's milk. In addition, its fat content is usually 50-60 percent higher (or more) than that of cow's milk. Although the butterfat content is usually 6-8 percent,* it can go much higher in the milk of some well-fed dairy buffaloes and in the milk of Swamp buffaloes (which are not nor-

---

*An analysis of 7,770 records of Nili/Ravi buffaloes in herds at the Pakistan Research Institute showed that average butterfat content was 6.40 (a mean based on 10 tests over 10 months). Of all the samples tested, 77 percent ranged between 5 and 8 percent butterfat and 12 percent were below 5 percent butterfat. (Information supplied by R. E. McDowell.)

26

mally used for milking). Cow's milk butterfat content is usually between 3 and 5 percent.

Because of its high butterfat content, buffalo milk has considerably higher energy value than cow's milk. Phospholipids are lower but cholesterol and saturated fatty acids are higher in buffalo milk. Studies have shown that digestibility is not adversely affected by this. Because of the high fat content, the buffalo's total fat yield per lactation compares favorably with that of improved breeds of dairy cattle; it is much higher than that of indigenous cows.

Normally the protein in buffalo milk contains more casein and slightly more albumin and globulin than cow's milk. Several researchers have claimed that the biological value of buffalo milk protein is higher than that of cow's milk, but this has not yet been proved conclusively.

TABLE 1 Comparative Gross Composition (in Percent) of Buffalo and Cow's Milk

	Total Solids	Fat	Protein	Lactose
River buffalo	17.96	7.45	4.36	4.83
Swamp buffalo	18.34	8.95	4.13	4.78
Friesian cow	12.15	3.60	3.25	4.60
Native cow	13.45	4.97	3.18	4.59

NOTE: In general, buffalo milk is considerably higher in total solids, fat, and protein than milk from improved or native cows. Buffalo milk is slightly higher in lactose.
Source: R. E. McDowell, Department of Animal Science, Cornell University.

TABLE 2 Some Physical Characteristics of Milk

Characteristics	Buffalo	Cow
Viscosity (cP)	2.04	1.86
Refractive index	1.3448	1.3338
Surface tension (dynes/cm)	55.4	55.9
pH	6.7	6.6
Freezing point depression	.560	.570
Average size of fat globules (μm)	5.01	3.85
Number of fat globules (millions/mm^3)	3.20	2.96
Phosphatase activity (units/100)	28	82
Fluorescence under UV light	Greenish yellow	Pale bluish

Source: H. Laxminarayana and N. N. Dastur. 1968. Buffaloes' milk and milk products. Part II. Dairy Science Abstracts 30:231–241.

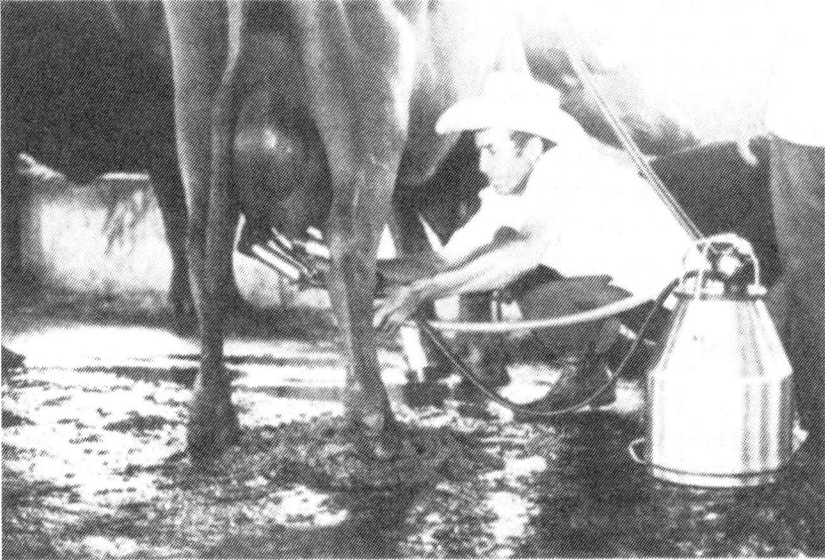

The milking buffalo. Left: City street, India. (I. MacRae, FAO Photo) Above: Private farm, São Paulo, Brazil. (E. Aziz Haik) Below: Cooperative dairy, Aarey, India. (W. R. Cockrill)

The mineral content of buffalo milk is nearly the same as that of cow's milk except for phosphorus, which occurs in roughly twice the amount in buffalo milk. Buffalo milk tends to be lower in salt.

Buffalo milk lacks the yellow pigment carotene, precursor for vitamin A, and its whiteness is frequently used to differentiate it from cow's milk in the market. Despite the absence of carotene, the vitamin A content in buffalo milk is almost as high as that of cow's milk.* (The two milks are similar in B-complex vitamins and vitamin C, but buffalo milk tends to be lower in riboflavin.)

Milk Products

Buffalo milk, like cow's milk, can provide a wide variety of products: butter, butter oil (clarified butter or ghee), soft and hard cheeses, condensed or evaporated milks, ice cream, yogurt, and buttermilk. It is of great economic importance in India in preparing "toned" milk—a mixture of buffalo milk and milk made by reconstituting skim milk powder.

The richness of buffalo milk makes it highly suitable for processing. To produce 1 kg of cheese, a cheesemaker requires 8 kg of cow's milk but only 5 kg of buffalo milk. To produce 1 kg of butter requires 14 kg of cow's milk but only 10 kg of buffalo milk. Because of these high yields, processors appreciate the value of buffalo milk.

Buffalo cheese is pure white. It many countries it is among the most desirable cheeses (mozzarella and ricotta in Italy, gemir in Iraq, the salty cheeses of Egypt, and pecorino in Bulgaria, for example). In Venezuela all the cheese produced from the small La Guanota milking herd in the Apure River basin (about 100 kg a day) is bought by the Hilton Hotel and sells for 15 bolivars per kg compared with 8 bolivars per kg for cheese made from cow's milk.

Although much in demand for making soft cheese, buffalo milk is less desirable for making hard cheeses such as cheddar or gouda. During cheese-making it produces acid more slowly than cow's milk, retains more water in the curd, and loses more fat in the whey.

Cheeses are becoming increasingly popular throughout the world. Demand is rising at a rate that is among the highest for any food product. Cheese offers particular benefit to areas where refrigeration is not widely available, where transporting high-protein foods to remote areas is difficult, and where seasonal fluctuations affect milk supplies. Buffalo milk may make cheese-making profitable on an even smaller scale than conventional dairying; it is more concentrated than cow's milk and requires relatively less energy to transport and process (an increasingly important factor where fuels are limited).

---

*Apparently the buffalo converts the carotene in its diet to vitamin A.

TABLE 3   Highest Milk Yield (kg per Day) Recorded in the All India Milk Yield
Competition

Year	Murrah	Other Buffaloes
1970–71	31.5	–
1971–72	25.8	–
1972–73	22.2	26.3
1973–74	23.8	29.1
1974–75	19.2	26.1
1975–76	24.2	27.5
1976–77	–	26.8

Source: Farm Information Unit, Directorate of Extension, Ministry of Agriculture
and Irrigation, New Delhi.

Yield

In countries like India and Egypt, the milk yield of buffaloes is generally higher (680–800 kg) than for local cattle (360–500 kg). However, since selection for exceptional milk production is not conducted systematically, large variations in yield occur between individual animals, and milk production of dairy buffaloes falls short of its potential.

Nonetheless, some outstanding yields have been recorded. On Indian government farms, average yields for milking buffaloes range from 4 to 7 kg per day in lactations averaging 285 days. Daily yields of 12 kg have been reported for some Bulgarian buffalo cows and a daily production of over 20 kg has been reported for some remarkable animals in India. A peak milk yield of 31.5 kg in a day has been recorded from a champion Murrah buffalo in the All India Milk Yield Competition conducted by the Government of India (see Table 3).

At Caserta, Italy, a herd of 1,600 machine-milked, pedigreed dairy buffaloes has produced average yields of 1,500 kg during lactations of 270 days. In Pakistan an analysis of over 6,000 lactations of Nili/Ravi buffalo cows showed an average yield of 1,925 kg during lactations averaging 282 days.* In India the average milk yield of Murrah buffaloes in established herds is also reported to be about 1,800 kg.†

Table 4 lists some outstanding lactation yields reported from different parts of the world.

---

*Average adjusted for year and season and calving. Cady et al., in press.
†Williamson and Payne, 1965.

As with cattle, the percentages of fat, protein, and total solids decrease as the milk yield increases.

The Swamp buffaloes of Southeast Asia are usually considered poor milk producers. They are used mainly as draft animals, but it may be that their milk potential has been underestimated. In the Philippines Swamp buffalo cows with nursing calves have produced 300–800 kg of milk during lactation periods of 180–300 days.* In Thailand Swamp buffaloes selected and reared for milk production have yielded 3–5 kg per day during 305-day lactations.†

TABLE 4   Milk Production of Some Outstanding Buffalo Cows and Dairy Herds

Location	Lactation Yield (kg)	Lactation Length (days)	Type
Salerno, Italy	3,498	–	Mediterranean dairy buffalo, single outstanding cow
Belém, Brazil	2,328	–	Mediterranean buffalo cow, Amazonian conditions[a]
Belém, Brazil	2,640	–	Crossbred Murrah/ Mediterranean buffalo cow, Amazonian conditions[a]
Punjab, India	2,200	–	River buffalo, average achieved on good farms[b]
	3,000	–	River buffalo, outstanding cows[b]
	4,000	305	Two River buffalo cows at Punjab Agricultural University[b]
Pakistan	4,300	285	Nili/Ravi cows, highest milk production from the records of over 6,000 lactations[c]
	5,337	408	Highest milk yield from individual cows in these 6,000 lactations[c]

NOTE: These figures, collected from various sources, are higher than those recorded for the average buffalo in the average Asian village. They indicate, however, the animal's potential as a milk animal and the opportunities for increasing milk production through selective breeding.

[a]Nascimento et al., 1979.
[b]Information supplied by D. Singh.
[c]Information supplied by R. E. McDowell.

*Philippines Council for Agriculture and Resources Research (PCARR). 1978. *The Philippines Recommends for Carabao Production*, PCARR, Los Baños, Philippines.
†Information supplied by Charan Chantalakhana.

The Nanning Livestock Research Institute and Farm in Kwangsi Province, which is representative of many others in South China, is upgrading the native Swamp buffaloes (or Shui Niu) by selective breeding for size and weight and by crossbreeding with dairy breeds such as the Murrah and Nili/Ravi. The crossbreeds that are milked yield 4–5 kg daily.*

Dairy Management

The characteristics of the dairy buffalo so closely approximate those of the dairy cow that successful methods of breeding, husbandry, and feeding for milk production for the cow can be applied equally to the dairy buffalo. Buffaloes, however, have not been bred for uniform udders and it is more difficult to milk them by machine.† Also, some buffaloes have more of a problem with milk letdown than dairy cows (although not as much of a problem as some native cattle breeds in the tropics). Frequently, a calf is kept with the cow and is tied to her foreleg at milking time. In India, Burma, and other countries a dummy calf may be provided; playing music seems to work, too.

Selected Readings

Addeo, F., Mercier, J. C., and Ribadeau-Dumas, B. 1977. The caseins of buffalo milk. *Journal of Dairy Research* 44(3):455–468.

Agarwala, O. P. 1962. Certain factors of reproduction and production in a water buffalo herd. *Indian Journal of Dairy Science* 15(2):45–51.

Albourco, F., Mincione, B., Addeo, F., and Ameno, M. 1969. Buffaloes' milk. II. Variations in composition during lactation of milk produced without change in feeding. *Industrie Agraria* 7(5):210–219.

Associazione Italiana Tecnici del Latter. 1970. Buffaloes' milk and cheese. *Scienza e Tecnica* 21(3):175–196.

Bhatnagar, V. K., Lohia, K. L., and Monga, O. P. 1961. Effect of the month of calving on milk yield, lactation length and calving interval in Murrah buffaloes. *Indian Journal of Dairy Science* 14(3):102-108.

Bhasin, N. R., and Desai, R. N. 1967. Effect of age at first calving and first lactation yield on life-time production in Hariana cattle. *Indian Veterinary Journal* 44:684–694.

Cady, R. A., Shah, S. K., Schermerhorn, E. C., and McDowell, R. E. In Press. Factors affecting performance of Nili-Ravi buffaloes in Pakistan. *Journal of Dairy Science.*

Castillo, L. S. 1975. Production, characteristics and processing of buffalo milk. In: *The Asiatic Water Buffalo.* Proceedings of an International Symposium held at Khon Kaen, Thailand, March 31–April 6, 1975. Food and Fertilizer Technology Center, Taipei, Taiwan.

Dassat, P. M., de Paolis, P., and Sartore, G. 1966. Environmental effects on milk yield in Italian buffalo. *Acta Medica Veterinaria* 12(6):587–593.

*Cockrill, W. R. 1976. *The Buffaloes of China.* FAO, Rome.
†Some thousands of buffaloes are machine milked in Bulgaria and Italy, however. At Ain Shams University in Egypt, buffaloes have adapted to machine milking. The calves are separated from their dams immediately after birth and no problems of milk letdown have been observed. (Information supplied by M. El Ashry.)

Dave, B. K., and Taylor, C. M. 1975. A study on relationship of persistency with other first production traits in Indian water buffaloes. *Indian Journal of Animal Health* 14(1):77–80.

Deshmukh, S. N., and Choudhury, P. N. R. 1971. Repeatability estimates of some economic characteristics in Italian buffaloes. *Zentralblatt fuer Veterinaer-medizen* 18:104–107.

Ganguli, N. C. 1979. Buffalo milk technology. *World Animal Review* 30:2–10.

Gomez, I. V. 1977. The manufacture of semi-hard cheese (Danish type) from carabao milk. *Philippine Agriculturalist* 61(3/4):78–86.

Gurnani, M., and Nagarcenkar, R. 1971. Evaluation of breeding performance of buffaloes and estimation of genetic and phenotypic parameters. *Annual Report of the National Dairy Research Institute*, Karnal, India.

Katpatal, B. G. 1977. Dairy cattle crossbreeding in India. I. Growth and development of crossbreeding. *World Animal Review* 22:15–21.

Kay, H. D. 1974. Milk and milk production. In: *The Husbandry and Health of the Domestic Buffalo*, edited by W. R. Cockrill. Food and Agriculture Organization of the United Nations, Rome, Italy.

Kohli, M. L., and Malik, D. D. 1960. Effect of service period on total milk production and lactation length in Murrah buffaloes. *Indian Journal of Dairy Science* 13(3): 105–111.

Nascimento, C. N. B., Moura Carvalho, L. O. D., and Lourenço, J. B. 1979. *Importância do Búfalo para a Pecuária Brasileira*. Agricultural Research Center for Humid Tropics (CPATU), Belém, Pará, Brazil.

Raafat, M. A., El-Sayed, Abou-Hussein, Abou-Raya, A. K., and El-Shirbiny, A. 1974. Some nutritional studies of colostrum and milk of cows and buffaloes. *Egyptian Journal of Animal Production* 14(1):137–148.

Ragab, M. T., Asker, A. A., and Kamal, T. H. 1958. The effect of age and seasonal calving on the composition of Egyptian buffalo milk. *Indian Journal of Dairy Science* 11(1):18–28.

Saudi Arabia Standards Institution. 1978. Raw milk. Saudi Arabia Standard 98.

Sebastian, J., Panthulu, P. C., and Bhimasena, M. 1971. Composition of Surti buffalo milk. *Annual Report of the National Dairy Research Institute*, Karnal, India.

Sekhon, G. S., and Gehlon, M. S. 1966. Repeatability estimates of some economic traits in the Murrah buffalo. *Ceylon Veterinary Journal* 14:18–22.

Singh, B. B., and Singh, B. P. 1971. Comparative study of lifetime economics of Hariana versus Murrah buffaloes. *Indian Veterinary Journal* 48:485–489.

Singh, R. P. 1966. A study of production up to ten years of age in buffaloes maintained at military farms. *Indian Veterinary Journal* 43:986–992.

Singh, S. B., and Desai, R. N. 1962. Production character of Bhadawari buffalo cow. *Indian Veterinary Journal* 39:332–343.

Venkayya, D., and Anantakrishnan, C. P. 1957. Influence of age at first calving on the performance of Murrah buffaloes. *Indian Journal of Dairy Science* 10(1):20–24.

Williamson, G., and Payne, W. J. A. 1965. *An Introduction to Animal Husbandry in the Tropics*. Longman, London, United Kingdom.

4 Work

The water buffalo is the classic work animal of Asia, an integral part of that continent's traditional village farming structure. Probably the most adaptable and versatile of all work animals, it is widely used to plow; level land; plant crops; puddle rice fields; cultivate field crops; pump water; haul carts, sleds, and shallow-draft boats; carry people; thresh grain; press sugar cane; haul logs; and much more. Even today, water buffaloes provide 20-30 percent of the farm power in South China, Thailand, Indonesia, Malaysia, Philippines, and Indochina.* Millions of peasants in the Far East, Middle East, and Near East have a draft buffalo. For them it is often the only method of farming food crops.

As fuel becomes scarce and expensive in these countries, the buffalo is being used more frequently as a draft animal. In 1979 water buffalo prices soared in rural Thailand because of the increased demand.

Although Asian farms have increasingly mechanized in the last 20 years, it has often proved difficult to persuade the farmer to replace his buffalo with a tractor since the buffalo produces free fertilizer and does not require diesel fuel. Now there is renewed official interest in draft power. Sri Lanka has recently opened up large new tracts of farmland in the Mahawali Valley, creating such a demand for work animals that buffalo shortages have become a national development problem. Indonesia's transmigration schemes are also handicapped by shortages of animal power.

For many small farmers the buffalo represents capital. It is often the major investment they have. Buffalo energy increases their productivity and allows them to diversify. Even small farms have work animals that, like the farmer himself, subsist off the farm. Tractors usually require at least four hectares for economical operation, which precludes their use on most peasant farms. Further, the infrastructure to maintain machinery is often not readily available.

---

*Figures provided by A. J. de Boer. In India water buffaloes contribute much less to farm power (6–12 percent); bullocks are more commonly used. In Pakistan buffaloes are little used for farm power (1–2 percent) but provide much of the road haulage. Papua New Guinea has no tradition of using any work animal, but villagers are increasingly using buffaloes for farm work and the government is employing Filipinos to train them.

The working buffalo. Above: Hauling cotton (about 650 kg per cart), Pakistan. (Brigadier Z. Khan) Below: Hauling logs, Turkey. (A. Defever, FAO Photo)

Above: Pumping water, Sind, Pakistan. (U.S. Agency for International Development)
Below: Puddling clay for bricks, near Kandy, Sri Lanka. (L. Drejarre, FAO Photo)

Upper left: Pressing sugarcane, Sabah, Malaysia. (Sabah Information Service) Lower left: Riding buffalo, Marajo, Brazil. (A. Bloch, Manchete) Right: Plowing stubble, People's Republic of China. (E. Schulthess)

Buffaloes are also used for hauling. Buffalo-drawn carts carry goods between villages where road surfaces are unfit for trucks. The animals easily
traverse ravines, streams, paddies, and narrow and rocky trails. In the cities
carts can compete economically with trucks where the road surface is unpre-
pared, where loading or unloading takes longer than the journey itself, or
where the loads are too small and distances too short to make trucking
economical. For road haulage buffaloes are generally shod: the shoes are
flat plates fitted to each hoof.

Capacity for Work

The water buffalo is a sturdy draft animal. Its body structure, especially
the distribution of body weight over the feet and legs, is an important advantage. Its large boxy hooves allow it to move in the soft mud of rice fields.
Moreover, the buffalo has very flexible pastern and fetlock joints in the lower
leg so that it can bend back its hooves and step over obstacles more easily
than cattle. This water-loving animal is particularly well adapted to paddy
farming because its legs withstand continual wet conditions better than mules
or oxen.*

Although buffaloes are preferred by farmers in the wet, often muddy lowlands of Asia, mules, horses, and cattle move more rapidly and are preferred
in the dryer areas.

Water buffaloes do not work quickly. They plod along at about 3 km per
hour. In most parts of Southeast Asia they are worked about 5 hours a day
and they may take 6–10 days to plow, harrow, and grade one hectare of rice
field. Their stamina and drawing power increase with body weight.

Because they have difficulty keeping cool in hot, humid weather (see next
chapter), it is necessary to let working buffaloes cool off, preferably in a
wallow, every 2 hours or so. Without this their body temperatures may rise to
dangerous levels.

A pair of 3-year-old buffaloes costs about the same as a small tractor in
Thailand. But many farmers raise their own calves and there is no investment
beyond labor. The "fuel" for the animals comes mainly from village pastures
and farm wastes such as crop stubble and sugarcane tops. Buffaloes have an
average working life of about 11 years, but some work to age 20.

---

*Australian animal scientists working in Bogor, Indonesia, found that the puddling effect
of buffalo hooves on the soil was critical for rice cultivation in the local soils. Tractors
produced fields so porous that they drained dry. (Information supplied by A. F. Gurnett-
Smith.) On one research station near Darwin, Australia, buffaloes were used to prevent
water draining from a dam. (Information supplied by D. G. Tulloch.)

Harness*

The yoke used on working buffalo in Asia has changed very little in the last 1,500 years. It is doubtful that a working buffalo can exert its full power with it. The hard wooden yoke presses on a very small area on top of the animal's neck, producing severe calluses, galls, and obvious discomfort. The harness tends to choke the animal as the straps under the neck tighten into the windpipe. Since the traditional hitch is usually higher than the buffalo's low center of gravity, the animal cannot pull efficiently. Considerably more pulling power and endurance can be obtained by improving the harness. The situation is not unlike that in Western agriculture in the twelfth century when the horse collar—one of the most important inventions of the Middle Ages—first appeared. Before that, horses were yoked like buffalo and the harness passed across their windpipes and choked them as they pulled. Use of the horse collar improved pulling efficiency and speeded the development of transportation and trade.

The curved yoke now universally used on water buffalo contacts an area of the neck that is only about 200 cm² (little more than half the size of this page). The entire load is pulled on this small area and causes the wood to dig into the flesh.

A horse collar is a padded leather device that encircles the animal's neck. One modified in Thailand for use on water buffalo (see page 43) had a contact area of 650 cm², more than three times that of the yoke it replaced. The collar's padding pressed against the animal's shoulders, not its neck, and therefore did not choke it. Attached to the collar were wooden hames with the traces for hitching the animal to a wagon or plow. In trials a buffalo pulled loads 24 percent heavier with the collar than with the yoke, and the horsepower it developed increased by 48 percent.†

Another potentially valuable harness is the breast strap, a set of broad leather straps that pass over the animal's neck and back. One breast strap modified for water buffalo use had a contact area of 620 cm², almost as much as that of the horse collar, and in trials the buffalo pulled a load 12 percent heavier than with a yoke and the horsepower it developed increased by almost 70 percent.‡

*This section is based on the work of panelist J. K. Garner in Thailand and Vietnam.
†These trials were conducted by J. K. Garner in Thailand in 1958. In the maximum-load test the yoked buffalo failed to move a load of 570 kg, but it moved a load of 640 kg when fitted with the horse collar. In an endurance test the yoked animal took 35 minutes to pull a load 550 m, but harnessed with the collar it took only 21 minutes.
‡The same animal used in the horse collar trials pulled 700 kg with the breast strap; in the endurance test it took 18.5 minutes to travel the 550-m distance.

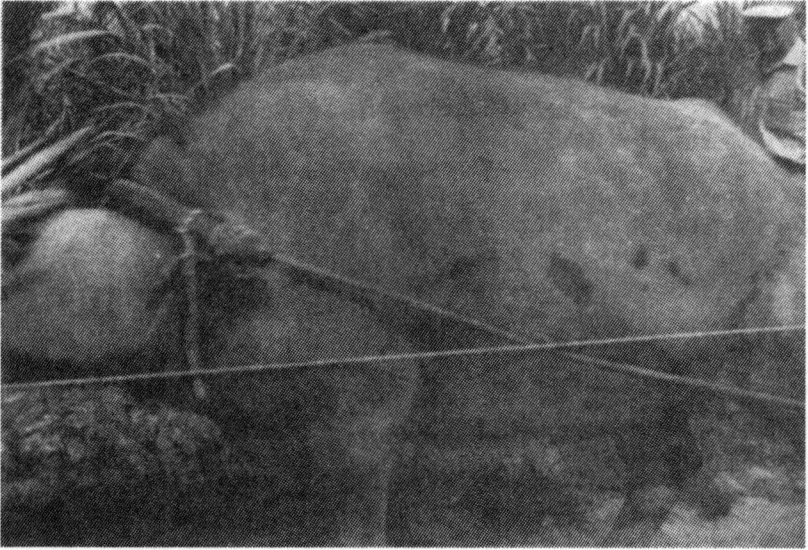

Traditional buffalo harness. Above: The wooden yoke presses on a small area of the neck, causing a large callus and probable discomfort. (J. K. Garner) Below: The strap around the lower neck digs into the animals' windpipe. (R. M. Fronda)

Experiments with alternate harness. Above: A leather horse collar. Below: A leather breast strap. With these harnesses this buffalo carries loads as much as 24 percent heavier than it could manage with the traditional wooden yoke. For details, see text. (J. K. Garner)

These seem very good innovations. In the humid tropics, however, leather collars and breast straps may decay rapidly. To make them widely practical may require experimentation with, or development of, special leather treatments or more durable materials.

Selected Readings

Cockrill, W. R. 1974. The working buffalo. In: *The Husbandry and Health of the Domestic Buffalo*, edited by W. R. Cockrill. Food and Agriculture Organization of the United Nations, Rome, Italy.

Cockrill, W. R. 1976. *The Buffaloes of China*. Food and Agriculture Organization of the United Nations, Rome, Italy.

de Guzman, M. R., Jr. 1975. The water buffalo—Asia's beast of burden and key to progress. In: *The Asiatic Water Buffalo*. Proceedings of an International Symposium held at Khon Kaen, Thailand, March 31–April 6, 1975. Food and Fertilizer Technology Center, Taipei, Taiwan.

Garner, J. K. 1958 (Reprint 1980) *Increasing the Work Efficiency of the Water Buffalo Through Use of Improved Harness*. (Copies available from Office of Agriculture, Development Support Bureau, Agency for International Development, Washington, D.C. 20523)

Kamal, T. H., Shehata, O., and Elbanna, I. M. 1972. *Isotope Studies on the Physiology of Domestic Animals*. International Atomic Energy Agency, Vienna, Austria.

Robinson, D. W. 1977. *Preliminary Observations on the Productivity of Working Buffalo in Indonesia*. Research Report No. 2, Centre for Animal Research and Development, Bogor, Indonesia.

Vaugh, M. 1945. Report on a detailed study of methods of yoking bullocks for agricultural work. *Indian Journal of Veterinary Science* 15:186–198.

Ward, G. M., Sutherland, T. M., and Sutherland, J. M. 1980. Animals as an energy source in Third World agriculture. *Science* 208:570.

5 Adaptability and Environmental Tolerance

Heat Tolerance

While the buffalo is remarkably versatile, it has less physiological adaptation to extremes of heat and cold than the various breeds of cattle. Body temperatures of buffaloes are actually lower than those of cattle, but buffalo skin is usually black and heat absorbent and only sparsely protected by hair. Also, buffalo skin has one-sixth the density of sweat glands that cattle skin has, so buffaloes dissipate heat poorly by sweating. If worked or driven excessively in the hot sun, a buffalo's body temperature, pulse rate, respiration rate, and general discomfort increase more quickly than those of cattle.* This is particularly true of young calves and pregnant females. During one trial in Egypt 2 hours' exposure to sun caused temperatures of buffalo to rise 1.3°C, whereas temperatures of cattle rose only 0.2–0.3°C.

Buffaloes prefer to cool off in a wallow rather than seek shade. They may wallow for up to 5 hours a day when temperatures and humidity are high. Immersed in water or mud, chewing with half-closed eyes, buffaloes are a picture of bliss.

In shade or in a wallow buffaloes cool off quickly, perhaps because a black skin rich in blood vessels conducts and radiates heat efficiently.† Nonetheless, wallowing is not essential. Experience in Australia, Trinidad, Florida, Malaysia, and elsewhere has shown that buffaloes grow normally without wallowing as long as adequate shade is available.

Cold Tolerance

Although generally associated with the humid tropics, buffaloes, as already noted, have been reared for centuries in temperate countries such as Italy, Greece, Yugoslavia, Bulgaria, Hungary, Romania, and in the Azerbaijan and

---

*Failure to appreciate this has caused many buffalo deaths in northern Australia when the animals were herded long distances through the heat of the day as if they were cattle.
†Tests at the University of Florida have shown that buffaloes in the shade cool off more quickly than cattle. (Robey, 1976)

45

Papua New Guinea

Australia

Although generally associated with Asia, water buffalo have shown remarkable adaptation elsewhere, such as in the Americas, parts of Europe, Australia, and Papua New Guinea. (Picture credits: Papua New Guinea, J. H. Schottler; Australia, Egypt, and U.S., N. D. Vietmeyer; Greece, A. Gagianas; Italy, P. Johnson, FAO; Yugoslavia, FAO photo; Tanzania, M. L. Kyomo; Brazil and Bolivia, H. Popenoe; Venezuela, J. A. Reggeti; Costa Rica, J. Luis Pacheco)

Greece

Italy

Yugoslavia

Tanzania

Egypt

United States

Brazil

Venezuela

Costa Rica

Bolivia

Georgian republics of the USSR. In 1807 Napoleon brought Italian buffaloes to the Landes region of southwest France and released them near Mont-de-Marsan. They became feral and multiplied prodigiously in the woods and dunes of the littoral, but unfortunately the local peasants found them easy targets, and with the fall of Napoleon the whole herd was killed for meat.* Buffaloes are also maintained on the high, snowy plateaus of Turkey as well as in Afghanistan and the northern mountains of Pakistan.

The buffalo has greater tolerance of cold weather than is commonly supposed. The current range of the buffalo extends as far north as 45° latitude in Romania and the sizable herds in Italy and the Soviet Union range over 40° N latitude.† Cold winds and rapid drops in temperatures, however, appear to have caused illness, pneumonia, and sometimes death. Most of the animals in Europe are of the Mediterranean breed, but other River-type buffaloes (mainly Murrahs from India) have been introduced to Bulgaria and the Soviet Union, which indicates that River breeds, at least, have some cold tolerance.

Altitude

Although water buffaloes are generally reared at low elevations, a herd of Swamp buffaloes is thriving at Kandep in Papua New Guinea, 2,500 m above sea level. And in Nepal, River buffaloes are routinely found at or above 2,800 m altitude.

Wetlands

Water buffaloes are well adapted to swamps and to areas subject to flooding. They are at home in the marshes of southern Iraq and of the Amazon, the tidal plains near Darwin, Australia, the Pontine Marshes in south-central Italy, the Orinoco Basin of Venezuela, and other areas.

In the Amazon, buffaloes (Mediterranean and Swamp breeds) are demonstrating their exceptional adaptability to flood areas. Buffalo productivity outstrips that of cattle, with males reaching 400 kg in 30 months on a diet of native grasses.‡

The advantage of water buffaloes over Holstein, Brown Swiss, and Criollo cattle was demonstrated in a test at Delta Amacuro, Venezuela, when the

---

*In the twelfth century Benedictine monks introduced buffaloes from their possessions in the Orient to work the lands of their abbey at Auge in northeastern France. In the thirteenth century a herd was introduced to England by the Earl of Cornwall, the brother of Henry III. Nothing is known about how well either herd survived.
†Philadelphia and Peking are at comparable latitudes. In the Southern Hemisphere the 40° line of latitude easily encompasses Cape Town, Buenos Aires, Melbourne, and most of New Zealand's North Island.
‡ Information supplied by C. Nascimento.

Kandep, Enga Province, Papua New Guinea, demonstrates that water buffaloes don't need lowland swamp sites. This area is more than 2,500 m above sea level, and for 4 years buffaloes there have lived on pitpit (*Saccharum* species), a coarse grass related to sugarcane. The buffalo have done well, without supplementary feeding. Cattle, on the other hand, were dying (possibly because of sodium deficiency) and had to be moved. (R. Nelson)

Buffalo are used as work animals even in winter in the Balkans. (Photo by Szakacz V. Sandor A.A.I., courtesy of István Csicsery-Rónay)

cattle developed serious foot rot in the wet conditions of the Orinoco Delta and had to be withdrawn from the test. The area of Venezuela is flooded 6 months of the year and creates constant problems for cattle, yet the buffalo seems to adapt well.*

High humidities seem to affect buffaloes less than cattle. In fact, if shade or wallows are available, buffaloes may be superior to cattle in humid areas.

In southern Brazil, trials comparing buffalo and cattle on subtropical riverine plains have favored the buffalo also. This work is being carried out on native pastures, mostly in the State of São Paulo.

Selected Readings

Hafez, E. S. E., Badreldin, A. L., and Shafei, M. M. 1955. Skin structure of Egyptian buffaloes and cattle with particular reference to sweat glands. *Journal of Agricultural Science, Cambridge* 46:19–30.

Katyega, P. M. J., Masoud, A. J., and Kobo, E. 1980. Information on body and carcass characteristics of Egyptian water buffalo and Mpwapwa x Friesian steers raised in central Tanzania. *Tanzania Veterinary Bulletin* 2:86–90.

Mason, I. L. 1974. Environmental physiology. In: *The Husbandry and Health of the Domestic Buffalo*, edited by W. R. Cockrill. Food and Agriculture Organization of the United Nations, Rome, Italy.

Moran, J. B. 1973. Heat tolerance of Brahman cross, buffalo, Banteng and Shorthorn steers during exposure to sun and as a result of exercise. *Australian Journal of Agricultural Research* 24(5):775–782.

Nair, P. G., and Benjamin, B. R. 1963. Studies on sweat glands in the Indian water buffalo. 1. Standardization of techniques and preliminary observations. *Indian Journal of Veterinary Science* 33:102.

Pandey, M. D., and Roy, A. 1969a. Studies on the adaptability of buffaloes to tropical climate. I. Seasonal changes in the water and electrolyte status of buffalo cows. *Indian Journal of Animal Science* 39:367.

Pandey, M. D., and Roy, A. 1969b. Studies on the adaptability of buffaloes to tropical climate. II. Seasonal changes in the body temperature, cardio-respiratory and hematological attributes in buffalo cows. *Indian Journal of Animal Science* 39:378.

Prusty, J. N. 1973. Role of the sweat glands in heat regulation in the Indian water buffalo. *Indian Journal of Animal Health* 12(1):33–37.

Robey, C. A., Jr. 1976. Physiological Responses of Water Buffalo to the Florida Environment. M.S. Thesis, University of Florida, Gainesville, Florida, USA.

*Information supplied by A. Ferrer.

6 Nutrition

Most buffaloes are located in countries where land, cultivated forage crops, and pastures are limited. Livestock must feed on poor-quality forages, sometimes supplemented with a little green fodder or by-products from food, grain, and oilseed processing. Usually feedstuffs are in such short supply that few animals have a balanced diet, but the buffalo seems to perform fairly well under such adverse conditions.

Growth Rate

Insufficient measurements have been taken to allow unequivocal statements about the relative growth rates of cattle and buffaloes. However, many observations made in various parts of the world indicate that the buffalo's growth is seldom inferior to that of cattle breeds found in the same environments. Some observations are given below.

- Trials in Trinidad in the early 1960s involved buffaloes grazing pangola grass (*Digitaria decumbens*) together with Brahman and Jamaican Red cattle. Over a period of 20 months the buffaloes gained an average of 0.72 kg per day, whereas the cattle on a comparable nearby pasture gained 0.63 per day.*
- In the Orinoco Delta of Venezuela unselected Criollo/Zebu crossbred cattle gained 0–0.2 kg per day on *Paspalum fasciculatum*, whereas the water buffaloes with them gained 0.25–0.4 kg per day.†
- In the Apure Valley of Venezuela, 100 buffalo steers studied in 1979 reached an average weight of 508 kg in 30 months, whereas the 30-month-old Zebu steers tested with them weighed 320 kg. The feed consumed was a blend of native grasses (25 percent of the diet) and improved grasses (such as pangola, para, and guinea grass).‡ In the same valley 200 buffalo heifers (air freighted from Australia) produced weight gains averaging 0.5 kg per day over

*Experiments performed by P. N. Wilson. (Information supplied by S. P. Bennett.)
†Cunha et al., 1975
‡Information supplied by A. Ferrer.

a 2-year period (and 72 percent of them calved). Government statistics for the area record average weight gains in crossbreeds between Zebu and Criollo cattle as 0.28 kg per day (with 40 percent calving).*

• In the Philippines, buffaloes showed weight gains of 0.75–1.25 kg per day, the same as those of cattle.†

• Daily weight gains of over 1 kg have been recorded for buffaloes in Bulgaria and Yugoslavia.‡

• Liveweight gains of 0.80 kg per day have been recorded for buffaloes in Papua New Guinea. In a very humid, swampy area of the Sepik River coastal plains liveweight gains by males averaged 0.47 kg per day and females 0.43 kg per day for more than a year. The average weight of 30 4-year-old female buffaloes was 375 kg, while the average weight of 4-year-old female Brahman/ Shorthorn crossbred cattle was 320 kg.§

• At the research station near Belém in the Brazilian Amazon weaned Murrah buffaloes, pastured continuously on *Echinochloa pyramidalis* (a nutritious grass), gained 0.8 kg daily and reached 450 kg in about 18 months.‖

• Liveweight gains of 0.74–1.1 kg per day have been obtained in Australia.** Buffalo steers grew as fast or faster than crossbred Brahman cattle on several improved pastures near Darwin,†† but on one very poor pasture, 4½-year-old buffaloes each weighed only 400 kg, whereas the Brahman crossbred steers reared with them weighed 500 kg.‡‡ The reason for this is not clear.

Efficiency of Digestion

Indian animal nutritionists have investigated water buffaloes intensively over the past two decades.§§ Many have reported that buffaloes digest feeds more efficiently than do cattle, particularly when feeds are of poor quality and are high in cellulose.‖‖ One trial revealed that the digestibility of wheat-straw cellulose was 24.3 percent for cattle and 30.7 percent for buffalo. The

*Information supplied by A. Ferrer.
†Information supplied by J. Madamba.
‡Information supplied by W. R. Cockrill.
§Information supplied by J. Schottler.
‖Nascimento, C. N. B., and Lourenço Junior, J. B., 1979. Criação de Búfalos na Amazônia. Belém, CPATU.
**Charles and Johnson, 1975; Ford and Parker, unpublished.
††Moran, 1973, unpublished.
‡‡Information supplied by D. G. Tulloch.
§§For reviews see Ichhponani et al., 1977; Ludri and Razdan, 1980.
‖‖Ichhponani et al., 1962; Ichhponani et al., 1971 a and b; Punj et al., 1968; Makkar and Sidhu, 1964; Mullick, 1964; Raghavan et al., 1963; Saini and Ray, 1964; Singh and Mudgal, 1967; Ludri and Razdan, 1980; Sahai et al., 1955.

Wherever buffaloes occur it is not uncommon to see skinny cattle and sleek buffaloes together. This originated the generally accepted notion that buffaloes utilize poor quality forage better than cattle do. (N. Vietmeyer, photographed near Darwin, Australia)

figures for berseem (*Trifolium alexandrinum*) cellulose were 34.6 percent for cattle and 52.2 percent for buffalo.* In another trial the digestion of straw fiber was 64.7 percent in cattle, 79.8 percent in buffalo.†

Other nutrients reported to be more highly digested in buffaloes than in cattle (Zebu) are crude fat,‡ calcium and phosphorus,§ and nonprotein nitrogen.||

Recent experiments in India suggest that buffaloes also are able to utilize nitrogen more efficiently than cattle. Buffaloes digested less crude protein than cattle in one trial but increased their body nitrogen more (and they were being fed only 40 percent of the recommended daily intake of crude protein).**

The ability of buffaloes to digest fiber efficiently may be partly due to the microorganisms in their rumen. Several Indian research teams have published data indicating that the microbes in the buffalo rumen convert feed into

*Sharma and Mudgal, 1966; Ichhponani and Sidhu, 1966.
†Sebastian et al., 1970.
‡Raghavan et al., 1963.
§Saini and Ray, 1964.
||Ichhponani and Sidhu, 1966; Ichhponani et al., 1962.
**Ludri and Razdan, 1980.

energy more efficiently than do those in cattle (as measured by the rate of production of volatile fatty acids in the rumen).*

Also, in laboratory studies samples of buffalo rumen contents produced volatile fatty acids more quickly from a variety of animal feedstuffs than did samples from the rumen of cattle.†

No single reason alone explains the buffalo's success in using poor quality forages. Rather, it is a combination of reasons that differ with the breed and conditions used. Studies by other researchers suggest that additional causes might include:

- Higher dry matter intake;‡
- Longer retention of feed in digestive tract;§
- Ruminal characteristics more favorable to ammonia-nitrogen utilization;||
- Less depression of cellulose digestion by soluble carbohydrates (e.g., starch or molasses);**
- Superior ability to handle stressful environments;†† and
- A wider range of grazing preferences.

Calf Growth Rates

Although the buffalo's gestation period is more than a month longer than that of cattle, the calves are born weighing 35–40 kg, or about the same as that of a newborn Holstein calf. But because buffalo milk has about twice the butterfat of cow's milk, the calves grow very quickly. They also suffer more shock at weaning and have to be slowly changed to their new feeding program. Buffaloes can be marketed as full-grown animals for beef at the age of 2–3 years, sometimes even earlier.

---

*Bhatia, 1967; Sidhu et al., 1967; Punj et al., 1968; Ichhponani et al., 1971a, and b; 1972; Verma et al., 1968; Langar et al., 1969; Tandon et al., 1972; Upadhaya et al., 1973.
†Ichhponani and Sidhu, 1965; Ichhponani et al., 1962.
‡Sahai et al., 1955; Upadhyaya et al., 1973.
§Ponappa et al., 1971. This is in dispute, however. A recent report concludes that although buffaloes will eat more low quality feed than cattle will it actually passes through more quickly and hence the buffalo's efficiency of feed utilization is no better than cattle's. Dement, T., Van Soest, P. J., 1981. Body size and herbivory. *Evolution Monograph.*
||Misra and Ranhotra, 1969; Nangia et al., 1972.
**Ichhoponani et al., 1969.
††In Papua New Guinea during the wet season cattle became distressed in the heat and humidity and ate less, whereas the buffaloes in the same field maintained appetite. The buffalo grazed in about four equal periods, which explains why they remained in better condition than cattle. (Information supplied by J. Schottler.)

TABLE 5   Some Coarse Grasses and Forages Palatable to Buffaloes but Less Readily Accepted by Cattle

Species	Common Name	Site of Observation
Cyperus species	sedges	Brazil
Eichhornia crassipes	water hyacinth	United States, Asia
Imperata cylindrica	cogon, blady grass, alang-alang, kunai	Papua New Guinea, United States
Pandanus species	screw pine	Australia
Paspalum conjugatum	cambute	Venezuela
Paspalum fasciculatum	chiguirera, mori	Venezuela, Brazil, Trinidad
Saccharum officinale	sugar cane (tops)	Asia, Trinidad
Saccharum species	pitpit	Papua New Guinea
Sida retusa	Paddy's lucerne	Australia
Typha angustifolia	cattail	United States
Zizaniopsis miliacea	giant cutgrass	United States

For example, in Indonesia it has been found that buffalo steers can be marketed 6 months before Zebu steers because they may be 100 kg heavier.* In Egypt some buffalo calves given feed supplemented with concentrates weighed 360 kg at 1 year of age.† At Ain Shams University near Cairo, buffalo calves weaned at 7-14 days of age gained 0.7 kg per day from weaning to slaughter at 18 months of age and weighed 400 kg. Rice straw comprised 50 percent of the finishing diet.

Buffaloes on Italian farms have reached 350 kg in 15-18 months and some year-old calves weighed 320 kg.‡

Grazing trials on native pasture (with mineral supplementation) in the Brazilian Amazon indicate that buffalo calves grew faster than cattle. At 2 years of age Mediterranean-type buffaloes averaged 369 kg, Swamp type, 322 kg, and Jafarabadi type, 308 kg. The Zebu cattle tested with them averaged 265 kg and the crossbred Zebu/Charolais, 282 kg.§

*Information supplied by M. Ottley.
†Information supplied by M. El Ashry.
‡Information supplied by W. R. Cockrill.
§Nascimento et al., 1979.

Water hyacinth, Philippines

Buffaloes living on poor forage. (Picture credits: Imperata grasses, Australia, D. G Tulloch; others N. D. Vietmeyer)

Sedges, Thailand

Swamp shrubs, United States

Feed Preferences

Buffaloes graze a wider range of plants than cattle. During floods near Manaus in the Brazilian Amazon when cattle become marooned on small patches of high ground, many suffer from foot rot and many starve to death. Their buffalo companions on the other hand—bodies sleek and full—swim out to islands of floating aquatic plants and eat them, treading water. Also, they dive almost 2 m to graze beneath the floodwaters.*

University of Florida buffaloes in a lakefront voluntarily consumed vines, sedges, rushes, floating aquatic weeds, and the leaves and shoots of willows and other trees along the water's edge. Few of these plants are voluntarily grazed by cattle. In northern Australia water buffaloes graze the very prickly leaves of pandanus; they also graze sedges, reeds, floating grass, and aquatic weeds.

Hungry buffaloes will eat bark, twigs, and other unpalatable vegetation. Because of the variety of their tastes they have been used in northern Queensland, Australia, to clear pastures of woody weeds left untouched by cattle. In some countries cattle are used to graze the palatable tops of pasture plants and are followed by buffaloes to graze the less desirable lower parts.

Selected Readings

Afifi, Y. A., El-Koussy, H. A., El-Khishen, S. S., and El-Ashry, M. A. 1977. Production of meat from Egyptian buffaloes. I. The effect of using different sources of roughages on rate of gain, gross and net efficiency during the growth and finishing periods. *Egyptian Journal of Animal Production* 17(2).

Appleton, D. C., Dryden, G., and Kondos, A. C. 1976. A comparison of the digestive efficiency of the water buffalo and the Brahman and Banteng cattle. *Proceedings of the Australian Society of Animal Production* 11.

Bhatia, I. S. 1967. *The Study of Factors Affecting the Utilization of Low Grade Roughages and Production of Volatile Fatty Acids in the Rumen of Indian Cattle.* Punjab Agricultural University, Ludhiana, India.

Chalmers, M. I. 1974. Nutrition. In: *The Husbandry and Health of the Domestic Buffalo,* edited by W. R. Cockrill. Food and Agriculture Organization of the United Nations, Rome, Italy.

Charles, D. D., and Johnson, E. R. 1975. Live weight gains and carcass composition of buffalo (*Bubalus bubalis*) steers on four feeding regimes. *Australian Journal of Agricultural Research* 26:407–413.

Chaturvedi, M. L., Singh, U. B., and Ranjhan, S. K. 1973. Comparative studies on the efficiency of feed ulitization in cattle and buffaloes. *Indian Journal of Animal Science* 43(12):1034.

*Information supplied by W. Kerr.

Chutikul, K. 1975. Ruminant (Buffalo) nutrition. In: *The Asiatic Water Buffalo*. Proceedings of an International Symposium held at Khon Kaen, Thailand, March 31–April 6, 1975. Food and Fertilizer Technology Center, Taipei, Taiwan.

Cunha, E., Alvarez, F., Larez, O., and Bryan, W. B. 1975. *Pasture and Livestock Investigations in the Humid Tropics: Orinoco Delta, Venezuela. 4. Beef Cattle and Water Buffalo Grazing Trials with Native and Introduced Grasses*. IRI Research Institute, New York, New York, USA.

Davendra, C. 1979. The potential value of grasses and crop by-products for feeding buffaloes in Asia. *Extension Bulletin* No. 126. Food and Fertilizer Technology Center, Taipei, Taiwan.

de Guzman, M. R., Jr. 1975. A summary of research into buffalo feeding in the Philippines. In: *The Asiatic Water Buffalo*. Proceedings of an International Symposium held at Khon Kaen, Thailand, March 31–April 6, 1975. Food and Fertilizer Technology Center, Taipei, Taiwan.

Dunkel, R. 1981. Animal production using nutrient-deficient fodders in the tropics and subtropics. *Animal Research and Development* 13:32–39.

El-Ashry, M. A., Mogawer, H. H., and Kishin, S. S. 1972. Comparative study of meat production from cattle and buffalo male calves. I. Effect of roughage: concentrate ratio in ration for rate of gain and feed efficiency of native cattle, Friesian and buffalo calves. *Egyptian Journal of Animal Production* 12(2):99–107.

Grant, R. J., Van Soest, P. J., McDowell, R. E., and Perez, C. B. 1974. Intake, digestibility and metabolic loss of Napier grass by cattle and buffaloes when fed wilted, chopped and whole. *Journal of Animal Science* 39(2):423.

Ibrovic, M., Cvetkovic, A., Smrcek, A., and Janevski, K. 1972. A contribution to the knowledge of metabolism of carotene and seasonal dynamics of vitamin A in the buffalo. *Veterinaria* 21(4):511.

Ichhponani, J. S., Makkar, G. S., Sidhu, G. S., and Moxin, A. L. 1962. Cellulose digestion in water buffaloes and Zebu cattle. *Journal of Animal Science* 21:1001.

Ichhponani, J. S., and Sidhu, G. S. 1966. Effect of urea on the voluntary intake of wheat straw in Zebu cattle and the buffalo. *Indian Veterinary Journal* 43:880–886.

Ichhponani, J. S., Makkar, G. S., and Sidhu, G. S. 1969. Biochemical processes in the rumen. III. Effect of different carbohydrates on *in vitro* digestion of cellulose and utilization of urea nitrogen by rumen micro-organisms from zebu and buffalo. *Indian Journal of Animal Science* 39:27–32.

Ichhponani, J. S., Makkar, G. S., and Sidhu, G. S. 1971a. Studies on the biochemical process in the rumen. VI. *In vivo* digestion of cellulose in buffalo and cattle. *Indian Veterinary Journal* 48:267–271.

Ichhponani, J. S., Makkar, G. S., and Sidhu, G. S. 1971b. Studies on the biochemical process in the rumen. VII. *In vitro* digestion of cellulose in cattle and buffalo. *Indian Veterinary Journal* 48:583–586.

Ichhponani, J. S., Makkar, G. S., and Sidhu, G. S. 1972. Studies on the biochemical processes in the rumen. VIII. The significance of molar proportions of acetic and propionic acids in the rumen on the growth rate of buffalo and Zebu. *Indian Veterinary Journal* 49:995–1000.

Ichhponani, J. S., Gill, R. S., Makkar, G. S., and Ranjan, S. K. 1977. Work done on buffalo nutrition in India—a review. *Indian Journal of Dairy Science* 30(3):173–191.

Langar, P. N., Sidhu, G. S., and Bhatia, I. S. 1969. Utilization of nitrogen and production of volatile fatty acids in the rumina of buffalo (*Bos bubalis*) and zebu (*Bos indicus*) at different time intervals, under a feeding regimen deficient in carbohydrates. *Indian Veterinary Journal* 46:593–598.

Ludri, R. S., and Razdan, M. N. 1975. Effect of level and source of dietary nitrogen on water metabolism in cows and buffaloes. *Indian Journal of Dairy Science* 28:107–109.

Ludri, R. S., and Razdan, M. N. 1980. Efficiency of nitrogen utilization by Zebu cows and buffaloes. I. Nutrient utilization and nitrogen balances on preformed protein diets. *Tropical Agriculture (Trinidad)* 57:83–90.

Makkar, G. S., and Sidhu, G. S. 1964. The fractionation of carbohydrates in Indian feed-stuffs and the digestion of different fractions in the rumina of cattle and buffalo. *Journal of Research, Ludhiana* 1:56–66.

Misra, R. K., and Ranhotra, G. S. 1969. Influence of energy levels on the utilization of peanut protein-urea nitrogen by cattle and buffalo. *Journal of Animal Science* 28: 107–109.

Mullick, D. N. 1964. Review of the investigations on the physiology of Indian buffaloes. *Indian Journal of Dairy Science* 17:45–50.

Naga, M. A., El-Shazly, K., Deif, H. I., Abaza, M. A., and El-Faham, R. H. 1975. Relationship between nitrogen balance, digested nitrogen and dry matter digestibility in ruminants. *Journal of Animal Science* 40(2):366–373.

Nangia, O. P., Aggarwal, V. K., and Singh, A. 1972. Studies on the utilization of dietary protein in cattle and buffaloes. *Indian Journal of Dairy Science* 25:1–5.

Nascimento, C. N. B., Moura Carvalho, L. O. D., and Lourenço, J. B. 1979. *Importância do Búfalo para a Pecuária Brasileira*. Agricultural Research Center for Humid Tropics (CPATU), Belém, Pará, Brazil.

Nascimento, C. N. B., and Velga, J. B. 1973. Weight gains in stabled buffaloes of the Mediterranean breed. *Boletim Tecnico do Instituto de Pesquisas e Experimentacao Agropecuarias do Norte* 58:24–72.

Ponnappa, G. G., Nooruddin, M. D., and Raghavan, G. V. 1971. Rate of the passage of food and its relation to digestibility of nutrients in Murrah buffaloes and Hariana cattle. *Indian Journal of Animal Science* 41:1026–1030.

Punj, M. L., Kochar, A. S., Bhatia, I. S., and Sidhu, G. S. 1968. *In vitro* studies on the cellulolytic activity and production of volatile fatty acids by the inocula obtained from the rumen of Zebu cattle and Murrah buffalo on different feeding regimens. *Indian Journal of Veterinary Science* 38:325–332.

Raghavan, G. V., Kakkar, B., Rao, M. V. N., and Mullick, D. N. 1963. Effects of air temperature and humidity on the metabolism of food nutrients in cattle and buffalo bulls. *Annals of Biochemistry and Experimental Medicine* 23:23–28.

Ray, S. N., and Mudgal, V. D. 1962. Rumen metabolism studies in cattle and buffalo. *Proceedings of the 16th International Dairy Congress, Copenhagen*. A:105–111.

Sahai, L., Johri, P. N., and Kehar, N. D. 1955. Effect of feeding alkali and water-treated cereal straws on milk yield. *Indian Journal of Veterinary Science* 25:201–212.

Saini, B. S., and Ray, S. N. 1964. Comparative utilization of coarse fodders in cattle and buffalo. *Annual Report of the National Dairy Research Institute*, Karnal, India.

Schottler, J. H., Boromana, A., and Williams, W. T. 1977. Comparative performance of cattle and buffalo on the Sepik Plains, Papua New Guinea. *Australian Journal of Experimental Agriculture and Animal Husbandry* 17(87):550–554.

Sebastian, L., Mudgal, V. D., and Nair, P. G. 1970. Comparative efficiency of milk production by Sahiwal cattle and Murrah buffalo. *Journal of Animal Science* 30:253–256.

Sharma, C. B., and Mugdal, V. D. 1966. Studies on the lignin and cellulose contents of fodder crops and effect of lignification of cellulose digestion. *Indian Journal of Dairy Science* 19:100.

Sidhu, G. S., Kochar, A. S., and Makkar, G. S. 1967. Effect of supplementing Bajra stalks with lucerne and concentrates and their part replacement by urea on the production of volatile fatty acids in the rumen of cows and buffaloes. *Journal of Research, Ludhiana* 4:104–110.

Singh, B. K., and Mudgal, V. D. 1967. The comparative utilization of feed nutrients from lucerne hay in buffalo and crossbred zebu heifers. *Indian Journal of Dairy Science* 20:142–145.

Tandon, R. N., Sharma, D. D., and Mugdal, V. D. 1972. Effect of feeding urea with different levels of energy on the biochemical changes in the rumen of cows and buffaloes. *Indian Journal of Animal Science* 42:174–179.

Upadhyaya, R. S., Singh, U. B., and Ranjhan, S. K. 1973. Digestibility of nutrients and VFA concentration in zebu cattle and buffalo calves fed on green cowpea and maize. *Indian Journal of Animal Science* 43:583–588.

Verma, M. L., Sidhu, G. S., and Kochar, A. S. 1968. Effect of frequency on the production of VFA in the rumen by zebu cattle and buffalo. *Journal of Research, Ludhiana* 5(3):1–6.

Verma, M. L., Singh, N., Sidhu, G. S., Kochar, A. S., and Bhatia, I. S. 1970. The *in vitro* cellulose digestion and VFA production from some of the common Indian feeds using rumen inocula from Zebu cattle and buffalo. *Indian Journal of Dairy Science* 23:155–160.

7 Health

When compared with other domestic livestock, the water buffalo generally is a healthy animal. This is particularly impressive because most of them live in hot, humid regions that are conducive to disease, and the buffalo is a bovine susceptible to most diseases and parasites that afflict cattle. Although the reasons are not specifically known, the effect of disease on the buffalo and its productivity is often less deleterious than on cattle.

Antibiotics and vaccines developed for cattle work equally well on buffaloes. As a result, treatments are available for most of the serious diseases of buffaloes, although some are not very effective for either animal.

The greatest buffalo losses are often among calves. Newborn buffalo calves, like bovine calves, can succumb in large numbers to viruses, bacteria, and poor nutrition. This is largely due to poor management during the calf's first 2 months of life. For example, as noted previously, villagers in some countries often sell the valuable buffalo milk, thus depriving the calves.

Buffalo calf losses are often similar to those of the cattle around them, but the animal's proclivity for wallowing exposes calves to waterborne diseases. Further, a young one occasionally drowns when an adult rolls on top of it.

Reactions to some specific diseases and parasites are discussed below.

Pasteurellosis   Probably the water buffalo's most serious disease, pasteurellosis, or hemorrhagic septicemia, is caused by the bacterium *Pasteurella multocida* (*P. septica*). Buffaloes are more susceptible to it than cattle and die in large numbers where pasteurellosis occurs. A vaccine against pasteurellosis is effective in protecting both buffaloes and cattle; it is cheap and easily made.

Tuberculosis   Despite some claims to the contrary, the water buffalo is susceptible to the bovine strain of tuberculosis (*Mycobacterium bovis*). Scattered reports from different parts of India indicate no difference in the incidence of infection between cattle and buffaloes.* Other strains of mycobacteria have

---

*Information supplied by S. K. Misra, Department of Veterinary Medicine, Punjab Agricultural University, Ludhiana, Punjab, India.

been isolated from feral buffaloes and cattle in northern Australia but seem to have little effect on the animals.* Tuberculosis occurs among the buffalo herds of the world only because most are kept under unsanitary conditions.†

Brucellosis Buffaloes and cattle are equally susceptible to brucellosis. Although seldom reported as a problem elsewhere, brucellosis in Venezuela is increasing more rapidly among buffaloes than among cattle.‡ In India the disease is no more prevalent among buffaloes than among cattle.§ As many as 57 percent of some Venezuelan herds are infected with the disease. It is a frequent cause of abortion in buffaloes. Serologic procedures and measures developed for the control of the disease in cattle are also effective means of curbing this infection in buffaloes. (Consumption of raw milk or contact with aborted fetuses may cause undulant fever in humans.)

Mastitis Among milking buffaloes mastitis is a problem as it is in dairy cows, but to a lesser extent. It is likely to increase, however, as the milk production per individual buffalo is increased. The bacteria that cause mastitis in the buffalo are similar to those in cattle. Treatment and control programs used for cattle are equally effective for buffaloes.

Other Diseases Among the epizootic diseases, rinderpest and piroplasmosis seem to affect buffaloes as much as cattle. Foot-and-mouth disease also affects buffaloes, but to a lesser degree than cattle, producing smaller lesions and having a lower incidence. In northern Australia buffaloes deliberately infected with bovine pleuropneumonia bacteria exhibited slight fever, but the disease never appeared. No naturally occurring cases have been reported in buffaloes.||

Ticks Buffaloes are notably resistant, although not immune, to ticks. In a tick-infested area of northern Australia only 2 engorged female ticks were

---

*Information supplied by D. G. Tulloch.
†In 1905 buffaloes were introduced to Trinidad because the cattle herds (Zebu and Brahman breeds) were infected with tuberculosis and, in those days, it was thought that buffaloes were resistant to the disease. Most were housed in muddy, ill-kept pens and forced to eat sugarcane tops off the ground; consequently, in 1949 over 30 percent of the buffaloes reacted to the tuberculin test. (The cattle herds had 80 percent reactors.) Tuberculosis was eliminated in Trinidad's buffaloes by improving the sanitary conditions: installing concrete floors and mangers and cleaning the pens regularly. This, together with regular tuberculin testing and removal of reactors, led to such a dramatic improvement that buffalo tuberculosis is now virtually unknown. (Information supplied by S. K. Bennett.)
‡Information supplied by A. Ferrer.
§Information supplied by S. K. Misra.
||Information supplied by J. Whittem.

found on 13 adult buffaloes during a 2-year test.* Accordingly, healthy buffaloes are not commonly affected by diseases borne by ticks nor are the hides damaged by their bites. Since ticks are rarely found on buffaloes, anaplasmosis, theileriasis, and babesiosis, which are tick-borne, have little effect on buffaloes in the field. (Buffaloes and cattle are equally susceptible, however, if inoculated with East Coast fever, a form of theileriasis.) This is important because tick infestations in cattle are particularly troublesome in the tropics and the pesticides used to control them are becoming ineffective as the ticks develop resistance. The pesticides are also becoming expensive.

The basis of the buffalo's tick resistance is not known, but wallowing and

TABLE 6   Some Infections and Parasites of Buffalo

Disease	Cause	Status of Control Technology	Susceptibility Compared with Cattle[a]
Brucellosis	Bacterium	Vaccines available	Equally susceptible
Leptospirosis	Bacterium	Vaccine and chemotherapy	Probably equally susceptible and higher incidences
Pasteurellosis	Bacterium	Vaccine available	Equal or greater
Tuberculosis	Bacterium	Improved vaccine needed	Equal
Blackleg	Bacterium	Vaccine available	Greater (Thailand); probably equal in other areas
Anthrax	Bacterium	Vaccine available	Equal
Salmonellosis	Bacterium	Chemotherapy	Probably equal
Rinderpest	Virus	Vaccine available	Equal or greater (Iran)
Foot and mouth disease	Virus	Vaccine needs improvement	Equal, but generally less severe
Bovine viral diarrhea	Virus	Need improved diagnostic tests and vaccines	Probably equal
Rabies	Virus	Vaccine, canid and bat control	Probably equal, but incidence much less
Bluetongue	Virus	Vaccine investigation necessary	Not known
Buffalo pox	Virus	Vaccine investigation necessary	Peculiar to the buffalo, which is also susceptible to cowpox
Ephemeral fever	Virus	Vaccine needed	Not known

[a] These estimates are necessarily generalizations that may vary for different sites.

*Information supplied by M. Ottley.

rubbing may play a role in it; animals kept in experimental concrete pens in Australia have developed heavy tick infestation.*

Screwworm   Larvae of the screwworm fly (*Callitroga* species), a major pest of livestock in Central and South America and some other tropical areas, do not affect adult water buffalo. In Venezuelan areas where cattle (Zebu type) are severely infested, adult water buffaloes are virtually free of screwworm larvae and the umbilicus of newborn calves seldom if ever becomes infected.†

TABLE 6   Some Infections and Parasites of Buffalo, continued

Disease	Cause	Status of Control Technology	Susceptibility Compared with Cattle[a]
Malignant catarrhal fever	Virus	Investigation needed	Not known
Contagious pleuropneumonia	Mycoplasma	Vaccine and diagnostic tests need improvement	Less (Australia); doubtful if buffalo ever suffer natural infection
Anaplasmosis	Rickettsia-like agent	Improved vaccine needed	Less susceptible (India)
Sporadic bovine encephalomyelitis	Chlamydia	Investigation needed	Not known
Trypanosomiasis	Protozoa	Vector control; new control measures needed	Equally susceptible
Theileriosis	Protozoa	Vector control; new control measures needed	Considered to be equally susceptible in Africa
Babesiosis	Protozoa	Vector control; new control measures needed	Less (Australia); clinical babesiosis rare
Coccidiosis	Protozoa	Effective chemotherapy	Equally susceptible
Sarcosporidiosis	Protozoa	Sanitary measures	Probably higher incidence
Echinococcosis	Helminth	Sanitary measures	Greater (Papua New Guinea); probably equally susceptible elsewhere
Schistosomiasis	Helminth	Snail control	Probably equally susceptible
Fascioliasis	Helminth	Snail control	Higher incidence owing to greater access to cercariae

[a]These estimates are necessarily generalizations that may vary for different sites.

*Trials carried out on Magnetic Island by R. H. Wharton, Commonwealth Scientific and Industrial Research Organisation, Townsville, Queensland, Australia.
†Information supplied by A. Ferrer.

The same is true in Papua New Guinea.* It is thought that the mud plaster produced by wallowing suffocates the larvae, but in India screwworms do not affect water buffaloes either, and there they wallow in fairly clear water and the farmer usually washes them off.†

Roundworm   The heavy losses of young buffalo calves throughout the world are caused, in large measure, by the roundworm *Toxocara vitulorum.* The calves seem more susceptible than mature animals and they become infected before birth or within 24 hours after birth through the mother's colostrum. The roundworm is the most serious buffalo parasite and in untreated calves the small intestine can get packed with worms to the point of complete occlusion. Although huge numbers of calves die each year, anthelmintic drugs that control roundworms are highly effective and widely available.

The adult water buffalo appears to have a high degree of resistance to strongyloid nematodes. Being such excellent converters of rough fodders they do not suffer the nutritional deficiency and the resulting liability to these roundworms experienced seasonally by cattle.*

Liver Fluke   During wallowing, water buffaloes can easily become infected with the waterborne infective stages of liver fluke (*Fasciola gigantica*). Although the number of flukes in a buffalo may be phenomenally high, no clinical signs of the disease are usually evident. It seems likely that the resulting liver damage reduces the growth and the work and milk production of buffaloes more than is generally appreciated.

Trypanosomiasis   The water buffalo is susceptible to trypanosomiasis and is reportedly more susceptible than cattle to *Trypanosoma evansi.*† Experience with the animal in Africa is limited, but trypanosomiasis may be the reason why Egypt is the only African country that has traditionally employed water buffalo.

Other Parasites   The wallow and its resulting mud cake seem to protect water buffalo from many biting flies, but the main ectoparasite in Australia and Southeast Asia is the buffalo fly (*Siphona* spp.). Pediculosis, caused by the sucking louse (*Hematopinus tuberculatus*), occurs widely among buffalo, and sarcoptic mange (*Sarcoptes scabiei* var. *bubalus*) is a serious disease, especially among calves and during dry seasons when wallowing opportunities are restricted. The lung worm *Dictyocaulus viviparus* thrives in warm, humid areas and sometimes infects buffaloes heavily, although its outward manifestations are rare.

---

*Information supplied by J. Schottler.
†Information supplied by S. K. Misra.

Selected Readings

Barlow, J. N. 1977. Toxicology and safety evaluation of Phosvel to Egyptian water buffalo. Pages 517-522 in: *Pesticide Management and Insecticide Resistance*, edited by D. L. Watson and A. W. A. Brown. Academic Press, New York, New York, USA.

Bhowmik, M. K., and Iyer, P. K. R. 1977. Studies on the pathology of chronic lesions in the mammary glands of buffaloes *(Bubalus bubalis)*. I. Actinomycosis, nocardiosis and lesions simulating tuberculosis. *Indian Veterinary Journal* 54(5):342-346.

Bhowmik, M. K., Singh, S. P., and Iyer, P. K. R. 1977. Studies on chronic mastisis in buffaloes *(Bubalus bubalis)*. *Indian Journal of Animal Health* 16(2):157-160.

Cockrill, W. R. 1974. Aspects of disease. In: *The Husbandry and Health of the Domestic Buffalo*, edited by W. R. Cockrill. Food and Agriculture Organization of the United Nations, Rome, Italy.

Dhar, S., Bhattacharyulu, Y., and Gautam, O. P. 1973. Susceptibility of Indian water buffalo *(Bubalus bubalis)* to Theileria annulata infection. *Haryana Agricultural University Journal of Research* 3(1):27-30.

Dwivedi, S. K., Mallick, K. P., and Malhotra, M. N. 1979. Babesiosis: clinical cases in Indian water buffaloes. *Indian Veterinary Journal* 56(4):333-335.

Griffiths, R. B. 1974. Parasites and parasitic diseases. In: *The Husbandry and Health of the Domestic Buffalo*, edited by W. R. Cockrill. Food and Agriculture Organization of the United Nations, Rome, Italy.

Hafez, S. M., and Krauss, H. 1979. Detection of antibodies against some respiratory pathogens in the sera of domestic animals in Egypt. *Bulletin of Animal Health and Production in Africa* 27(3):209-214.

Ho, T. M. 1976. Animal disease control program in Republic of China for better animal production. *Korean Journal of Animal Science* 18(4):317-321.

Iannelli, D. 1978. Water buffalo *(Bubalus bubalis* Arnee) allotypes: identification of a multiple allelic system. *Animal Blood Groups and Biochemical Genetics* 9(2):105-113.

Kapur, M. P., and Singh, R. P. 1977. Diagnosis of mastitis: a comparative study of four indirect tests. *Haryana Veterinarian* 16(2):69-73.

Leue, A. 1971. Buffaloes in Nepal. *Tierartliche Umschau* 26(4):173-178.

Mohan, R. N. 1968. Diseases and parasites of buffaloes. I. Viral, mycoplasmal, and rickettsial diseases. II. Bacterial and fungal diseases. III. Parasitic and miscellaneous diseases. *Veterinary Bulletin* 38:567-576, 647-659, 735-756.

Murthy, D. K., and Sharma, S. K. 1974. Studies on reactivity and immunogenecity of cell-culture rinderpest vaccine in different species of ruminants. *Indian Journal of Animal Science* 44(6):359-365.

San Agustin, F. 1973. Important diseases and parasites of Carabaos. *Philippine Journal of Nutrition* 26(2):68-79.

Schneider, C. R., Kitikoon, V., Sornmani, S., and Thiachantra, S. 1975. Mekong schistosomiasis. III. A parasitological survey of domestic water buffalo *(Bubalus bubalis)* on Khong Island, Laos. *Annals of Tropical Medicine and Parasitology* 69(2):227-232.

Sharma, K. N. S., Jain, D. K., and Noble, D. 1975. Calf mortality in pure and crossbred Zebu cattle and Murrah buffaloes reared artificially from birth. *Animal Production* 20(2):207-211.

Sharma, S. K., Banerjee, D. P., and Gautam, O. P. 1978. Anaplasma marginale infection in Indian water buffalo *(Bubalus bubalis)*. *Indian Journal of Animal Health* 17(2):105-110.

Shaw, J. J., and Lainson, R. 1972. Trypanosoma vivax in Brazil. *Annals of Tropical Medicine and Parasitology* 66(1):25-32.

Shukla, R. R., and Singh, G. 1972. Studies on tuberculosis amongst Indian buffaloes. *Indian Veterinary Journal* 49(2):119-123.

Soni, J. L. 1978. Suitability of different serological tests for diagnosis of brucellosis in buffaloes *(Bubalus bubalis)*. *Indian Journal of Animal Science* 48(12):873-881.

Tantawi, H. H., Fayed, A. A., Shalaby, M. A., and Skalinsky, E. I. 1979. Isolation, cultivation and characterization of poxviruses from Egyptian water buffaloes. *Journal of
  the Egyptian Veterinary Medical Association* 37(4):15-23.
Thomson, D. 1977. Diseases of water buffalo in the Northern Territory of Australia.
  Australian Veterinary Practioner 7(1):50-52.
Young, P. L. 1979. Infection of water buffalo (*Bubalus bubalis*) with bovine ephemeral
  fever virus. *Australian Veterinary Journal* 55(7):349-350.

8 Reproduction

The water buffalo has a reputation for being a sluggish breeder, but the average animal is so poorly fed that its reproductive performance is unrepresentative of its capabilities. Without reasonable nutrition the animals cannot reach puberty as early in life or reproduce as regularly as their physiology or genetic capability would normally allow.

Actually, adequately nourished buffaloes reach puberty at about the same age as cattle, as early as 18 months of age in buffalo bulls. In northern Australia Swamp females have conceived even as early as 14 months of age and feral buffaloes routinely conceive at 16 months of age.* In the herd at Punjab Agricultural University in Ludhiana, India, 11 River buffalo heifers showed estrus at ages less than 18.5 months and a few came into heat when less than 15 months old.†

The water buffalo also can calve at an age comparable to that of cattle. At the Ain Shams University in Egypt a well-fed Egyptian buffalo herd of several hundred animals has an average age at first calving of 27 months, 22 days.‡ Most animals in the Punjab Agricultural University River buffalo herd calved before 35 months, one at 28.3 months.† In one Venezuelan herd almost all heifers 20–24 months old were pregnant; virtually all calved before age 38 months, most by 30 months, and one at age 23 months.§

In trials in Queensland, Australia, and in Papua New Guinea buffaloes produced more calves over a 3-year period than the cattle tested with them. In the hot, humid Sepik Plains in northern Papua New Guinea it was noticed that female buffaloes (Swamp breed) came into estrus even while they were losing weight because of inadequate nutrition, whereas cattle did not. Under these stressful conditions the buffalo calves also reached sexual maturity

---

*Information supplied by D. G. Tulloch.
†Information supplied by D. Singh.
‡Information supplied by M. El Ashry. Because of nutritional uncertainties, El Ashry and his colleagues believe that body weight is a better indicator of sexual preparedness than age is. These researchers at Ain Shams University recommend mating heifers when they weigh 365 kg no matter what their age. Research at Punjab Agricultural University shows that buffalo heifers can be bred when they weigh over 270 kg and manifest estrus.
§Information supplied by A. Ferrer.

earlier and the buffaloes had a higher calving percentage and a shorter calving interval because they came back into estrus more quickly than cattle.* Similar observations have been made in Florida, Trinidad, the Brazilian Amazon, Venezuela, and elsewhere. Although these are exceptions to the normal observations in Asia, where buffaloes seem to breed more slowly than cattle, they do demonstrate the buffalo's potential for improved breeding.

Estrus in buffalo cows usually lasts about 24 hours, but duration varies and may range from 11 to 72 hours. It occurs on an average 21-day cycle. Determination of when a cow is in estrus is difficult because often the animal shows few outward signs of "heat." This increases the chances of missing a cycle, especially for artificial insemination. Unclean surroundings, poor nutrition, and poor management, cause a high death rate among calves; this also contributes to the buffalo's often low reproductive rate.

In many areas, calving is seasonal. This seems to be largely due to changes in nutrition. It may also be caused by heat stress, in either males or females, which results in a low breeding rate during the hot season. However, when buffalo cows are well fed, they come into estrus and will breed in any season.

Many matings take place at night and are therefore unobserved. In one set of pregnancy diagnoses in northern Australia, the buffalo's conception rate (81 percent) was higher than that of the Brahman crossbreeds (70 percent) they were with.† In India, artificial insemination of water buffaloes began in the late 1950s. Deep-frozen semen is now available and its use is spreading. Overall conception rates of 70–80 percent are obtained. It is estimated that some 100,000 buffaloes are now being artificially inseminated.‡

The water buffalo's gestation period is about one month longer and is more variable than that of cattle. Whereas cattle give birth after about 280 days (Angus, 279, Holstein, 279–280, Brown Swiss, 286), buffaloes take 300–334 days (average 310) or roughly 10 months and 10 days (differences between breeds are unknown).§ In Punjab, India, River buffaloes have been observed to come into estrus as early as 40 days after calving. ‖

Nonetheless, only under uncommon circumstances can a buffalo cow produce a calf each year. In one herd of 800 cows in Venezuela the average female buffalo over age 4 produces 2 calves every 3 years.** In response to a recent questionnaire, the majority of Indonesian farmers estimated that the calving rate was between 3 and 4 calves in 5 years. A few claimed a calf a

---

*Information supplied by J. Schottler. The age at first calving of more than 60 nutritionally poor buffaloes was 38 months in one herd and 45 months for Brahman cross cattle.
†Information supplied by D. G. Tulloch. Similar figures have been recorded in Venezuela (information supplied by A. Ferrer).
‡FAO, 1979.
§In Egypt, the average period, according to a study of 424 gestations, is 316 ± 8 days (Ghanem, 1955a and b).
‖Information supplied by D. Singh.
**Information supplied by A. Ferrer.

year, some only 1 or 2 calves in 5 years.* In Florida it has been noted that some buffalo cows having just calved became pregnant more quickly than cattle, so that a calf may indeed be produced each year.† Regular yearly breeding has been noted also in northern Australia.‡

The incidence of abortion, dystocia, retained placenta, and other parturition problems in buffaloes is similar to that in cattle. Twinning is very rare; probably no more than 0.01 percent of buffalo pregnancies produce twins.

Preliminary results in northern Australia indicate that weaning can be carried out as late as 12 months of age without any effect on conception time of the buffalo dam.‡

Selected Readings

Bhattacharya, P. 1974. Reproduction. In: *The Husbandry and Health of the Domestic Buffalo*, edited by W. R. Cockrill. Food and Agriculture Organization of the United Nations, Rome, Italy.

Chauhan, F. S., Singh, N., and Singh, M. 1977. Involution of uterus and cervix in buffaloes. *Indian Journal of Dairy Science* 30(4):286–291.

Eusibio, A. N. 1975. Breeding, management and feeding practices of buffaloes in the Philippines. In: *The Asiatic Water Buffalo*. Proceedings of an International Symposium held at Khon Kaen, Thailand, March 31–April 6, 1975. Food and Fertilizer Technology Center, Taipei, Taiwan.

FAO Animal Production and Health Paper. 1979. *Buffalo Reproduction and Artificial Insemination*. Proceedings of the Seminar Sponsored by FAO/SIDA/Government of India, held at National Dairy Research Institute, Karnal, 132001-India, December 4–15, 1978. Food and Agriculture Organization of the United Nations, Rome, Italy.

Ghanem, Y. S. 1955a. Environmental causes of variation in the length of gestation of buffaloes. *Indian Journal of Veterinary Science* 25:301–311.

Ghanem, Y. S. 1955b. Genetic causes of variation in the length of gestation of buffaloes. *Indian Journal of Veterinary Science* 25:307–311.

Health, E., and Gupta, R. 1976. Ultrastructure of water buffalo (*Bos bubalis*) spermatozoa. *Zentralblatt fuer Veterinaermedizin* 23(2):106–120.

Liu, C. H. 1978. The preliminary results of crossbreeding of buffaloes in China. Research Institute for Animal Science of Kwangsi, People's Republic of China.

Matharoo, J. S., and Singh, M. 1980. Revivability of buffalo spermatozoa after deep-freezing the semen using various extenders. *Zentralblatt fuer Veterinaermedizin* 27(5): 385–391.

Rao, A. V. N., Murthy, T. S., and Dutt, K. L. 1979. Effect of lunar cyclicity on oestrus rhythmicity in the Indian water buffalo. *Livestock Advisor* 4(7):29–30.

Rawal, C. V. S. 1978. A study on correlations of weight of pituitary gland and reproductive organs in males of Indian water buffalo. *Indian Journal of Heredity* 10(3):11–12.

Robinson, D. W. 1977. *Livestock in Indonesia*. Research Report No. 1. Centre for Animal Research and Development, Bogor, Indonesia. (In English and Indonesian.)

Roy, A. 1974. Observations on the physiology of reproduction. In: *The Husbandry and Health of the Domestic Buffalo*, edited by W. R. Cockrill. Food and Agriculture Organization of the United Nations, Rome, Italy.

---

*Robinson, 1977.
†Information supplied by H. Popenoe. Calving intervals of 14–15 months have been reported in Egypt and Venezuela (information supplied by M. El-Ashry and A. Ferrer).
‡Information supplied by D. G. Tulloch.

Sahai, R., and Singhal, R. A. 1977. The sex chromatin profile of water buffaloes of Murrah breed. *Indian Journal of Animal Science* 11(2):63–67.
Sharma, A. 1978. Studies on the process of spermatogenesis and epididymal sperm reserve in water buffalo (*Bubalus bubalis*). *Thesis Abstracts, Haryana Agricultural University* 4(4):331.
Singh, M., Matharoo, J. S., and Chauhan, F. S. 1980. Preliminary fertility results with frozen buffalo semen in Tris extender. *Theriogenology* 13(3):191–194.
Singh, N., Chauhan, F. S., and Singh, M. 1979. Postpartum ovarian activity and fertility in buffaloes. *Indian Journal of Dairy Science* 32(2):134–139.
Suntraporn Ratanadilok Na Puket. 1975. The improvement of buffalo production through breeding and management under Thailand conditions. In: *Asiatic Water Buffalo*. Proceedings of an International Symposium held at Khon Kaen, Thailand, March 31–April 6, 1975. Food and Fertilizer Technology Center, Taipei, Taiwan.
Takkar, P. O., Chauhan, F. S., Tiwana, M. S., and Singh, M. 1979. Breeding behaviour of buffalo cows. *Indian Veterinary Journal* 56:168–172.
Toelihere, M. R. 1975. Physiology of reproduction and artificial insemination of water buffaloes. In: *The Asiatic Water Buffalo*. Proceedings of an International Symposium held at Khon Kaen, Thailand, March 31–April 6, 1975. Food and Fertilizer Technology Center, Taipei, Taiwan.
Tulloch, D. G. 1968. Incidence of calving and birth weights of domesticated buffalo in the Northern Territory. *Proceedings of the Australian Society for Animal Production* 7:144–147.
Tulloch, D. G. 1979. The water buffalo in Australia: reproductive and parent-offspring behaviour of buffalo. *Australian Wildlife Research* 6:265–287.

9 Management

Water buffaloes are adaptable and are managed in many ways. In general, they are raised like cattle. But in some operations they must be handled differently. This chapter highlights these differences.

Millions of water buffaloes are managed in "backyards" in Asia. They exist on the resources of small holdings. Management and expenditures are minimal. Care of the family buffalo is usually entrusted to children, old people, or women not engaged in other farm duties; the buffalo allows them to be useful and productive.

The buffalo fits the resources available on the farm, but it is also an urban animal. Thousands of herds of 2-20 buffaloes may be found in the cities and towns of India, Pakistan, and Egypt—all fed, managed, and milked in the streets.

In addition, the buffalo has important qualities as a feedlot animal; it can be herded and handled with relative ease because of its placid nature. The Anand Cooperative in India's Gujarat State, which daily contributes thousands of gallons of milk to Operation Flood (the world's largest nutrition project), involves more than 150,000 Surti buffaloes that are fed, managed, and milked by their owners under feedlot-like conditions in their villages. Many of Italy's 100,000 buffaloes are maintained under similar conditions.

Water buffaloes can also be managed on rangelands. In Brazil, Venezuela, Trinidad, the United States, Australia, Papua New Guinea, Malaysia, Indonesia, the Philippines, and elsewhere there is rising interest in raising buffalo beef on the range. The production practices for raising them are similar to those used for range cattle.

Water buffaloes in the humid tropics must be able to cool off. Shade trees are desirable, and although a wallow is not essential, it is probably the most effective way the animal has of coping with heat. Alternatively, water showers may be provided to wet down the animals 3-5 times during the hottest part of the day.

Water buffaloes are intelligent animals. Young ones learn patterns quickly and often are reluctant to change their habits. Feral animals—even those born in the wild—tame down after a week or two in a fenced enclosure to the point

Traditional buffalo management, Punjab, India. (Agency for International Development)

Feedlot management of buffaloes, São Paulo, Brazil. (E. Aziz Haik)

where many can be handled, haltered, and hand fed.* Among feral herds of northern Australia it has been observed that buffaloes have clans and families. A female calf seems to remain with its family and mother for many years (possibly for life). A male calf stays until it is about 2 years old, when it is driven from the group by an adult bull.†

It has also been noted in northern Australia that free-ranging buffaloes instinctively select clean water areas to drink from, other areas to wallow in, and still others as "toilet areas." In addition, China's buffaloes reportedly are being "toilet trained" to defecate only at specific sites to avoid contaminating waterways with schistosome eggs.

Another interesting observation from northern Australia is that most buffalo dams readily adopt a calf that has been orphaned by the death of its mother.‡ In fact, females will allow several calves to nurse (including calves of other mothers and sometimes even adults).

Buffaloes are also self-reliant. For several months each year in Vietnam and Malaysia, for instance, they are turned loose in the forests to fend for themselves.

*In one example, 100 feral buffalo were captured in Northern Territory, Australia. Within 14 days all the animals—males and females, young and old—had become docile and amenable to handling. (Information supplied by D. G. Tulloch.)
†Information supplied by D. G. Tulloch.
‡Tulloch, 1979.

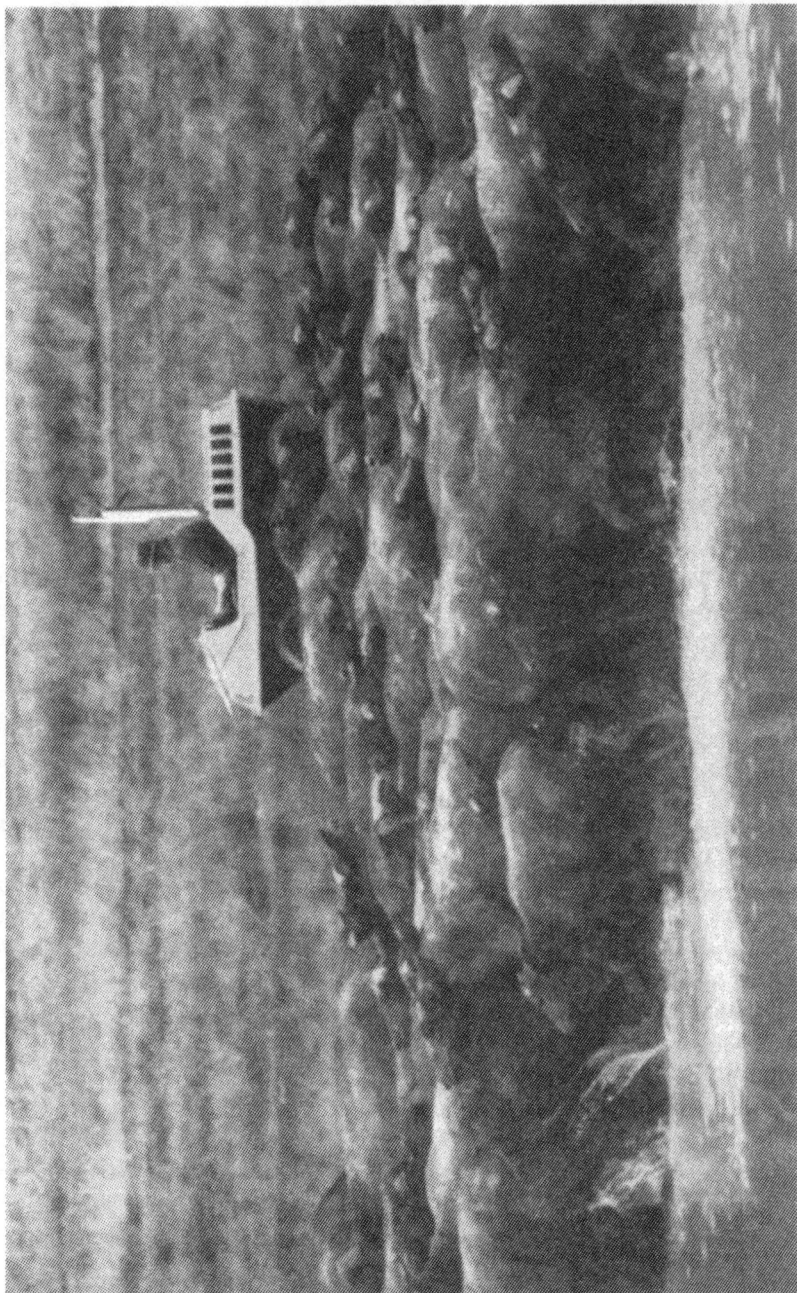

Range management of water buffaloes. Left: Near Darwin, Australia. (T. Nebbia © National Geographic Society) Above: Mudginberri, northern Australia. (D. Gordon) Below: Marajo Island, Brazil. (O. Cruzeiro magazine)

Provision of adequate fencing is one of the great problems of buffalo management. The animals have strong survival instincts and if feed runs short, such as in the dry season, they will break through fences that would deter cattle who would remain and starve. They will also break through fences if their family unit is split up. Barriers must be stronger than those used for cattle and the wires closer together and lower to the ground because buffaloes lift fences up with their horns rather than trample them down. In northern Australia, Papua New Guinea, and Costa Rica it has been found that buffaloes are particularly sensitive to electric fences (a single wire is all that is needed), and in Brazil a special suspension fence has been devised.* Both of these seem to be cheap and efficient answers to the fencing problem.

Water buffaloes are easily handled from horseback and easily worked through a corral. Actually, because of their docility they can be mustered on foot, even on ranges where cattle require horses. Unless they come from different areas they tend to herd together and can be mustered like sheep.

One of the major management adjustments to be made by cattlemen is understanding and capitalizing on the buffalo's placid nature. Buffaloes are naturally timid and startle easily; they must be handled quietly and calmly. Rough handling, wild riding, and loud shouting make handling them more difficult and training them much harder.†

Village buffaloes are led and managed by a ring threaded through the septum between the nostrils. The technique is frequently applied crudely and cruelly, often resulting in a ripped septum.

The identification of individual buffaloes is difficult. Fire brands do not remain legible on the skin for long. Cryobranding (freeze branding) is more durable. Most types of ear tags are not very successful; the numbers wear off and mud covers up the tag's color. In northern Australia ear tatooing has been the most successful identification technique, with tatoos remaining legible for at least 8 years.‡

When they pasture together, cattle and buffaloes coexist satisfactorily. They segregate themselves into their own groups and do not interfere with one another. The buffaloes, however, usually dominate the cattle and tend to monopolize the areas with the best feed supply.

---

*Moura Carvalho et al., 1979.

†Between 1958 and 1962 hundreds of Australian buffalo were shipped on the hoof to the meat markets in Hong Kong without trouble, despite crowded shipboard conditions and the long sea voyage. But in 1962 one roughly treated bull went beserk in Hong Kong and killed a handler, and the Hong Kong authorities stopped the trade as a result, although the problem was really one of mismanagement. Buffalo have since been sent by air from Australia to Venezuela, Nigeria, and Papua New Guinea and by barge to Papua New Guinea. No handling problems have been experienced en route.

‡Information supplied by D. G. Tulloch.

Feed troughs and mineral boxes used for cattle are suitable for buffaloes, but chutes and crushes must be widened to accommodate the buffalo's broader body and, when necessary, the Swamp buffalo's greater horn spread.

Water buffaloes are powerful swimmers. In Brazil they have been known to escape by swimming down the Amazon River. An unusual management difficulty is caused by piranha in the rivers and swamps of Venezuela. In one herd of 100 heifer buffaloes, 40 have lost all or part of a teat to these voracious fish.*

The horns of water buffaloes are seldom removed or prevented from growing, a testament to the animal's docility. (When questioned, one Thai villager said that he wouldn't allow it because it would be a disgrace to the buffalo.†) However, the animals can be dehorned as calves in the same manner as cattle. They are then easier to handle in chutes and cause less accidental injury to neighboring animals, handlers, walls, and trees.

Selected Readings

Benya, E. G. 1972. Cattle and buffalo production in Vietnam. Dairy Science Mimeo Report, Florida Agricultural Experiment Station, Gainesville, Florida, USA.

Cockrill, W. R. 1974. Management, conservation and use. In: *The Husbandry and Health of the Domestic Buffalo*, edited by W. R. Cockrill. Food and Agriculture Organization of the United Nations, Rome, Italy.

De Boer, A. J. 1976. A further report on the Taiwan livestock industry: general policy conclusions and further research needs. *Journal of Agricultural Economics* 19:116-129.

Lin, C. H. 1975. The breeding, management and feeding of water buffaloes in Taiwan. In: *The Asiatic Water Buffalo*. Proceedings of an International Symposium held at Khon Kaen, Thailand, March 31-April 6, 1975. Food and Fertilizer Technology Center, Taipei, Taiwan.

McDowell, R. E. 1972. *Improvement of Livestock Production in Warm Climates*. W. H. Freeman and Company, San Francisco, California, USA.

Moura Carvalho, L. O. D., Lourenço, J. B., Nascimento, C. N. B., and Costa, N. A. 1979. Cerca de Contencao para Bubalinos e Bovinos. *Comunicado Tecnico* No. 28, Agricultural Research Center for the Humid Tropics (CPATU), Belém, Pará, Brazil.

Rufener, W. H. 1975. Management and productive performance of water buffalo in northeast Thailand. In: *The Asiatic Water Buffalo*. Proceedings of an International Symposium held at Khon Kaen, Thailand, March 31-April 6, 1975. Food and Fertilizer Technology Center, Taipei, Taiwan.

Tulloch, D. G. 1978. The water buffalo in Australia: grouping and home range. *Australian Wildlife Research* 5:327-354.

Tulloch, D. G. 1979. Redomestication of water buffaloes in the Northern Territory of Australia. *Animal Regulation Studies* 2(1):5-20.

*Information supplied by A. Ferrer.
†Information supplied by Charan Chantalakhana.

10 Environmental Effects

The grazing and wallowing habits of water buffaloes may have unexpected consequences when the animals are introduced to new, perhaps fragile environments. The presence of several thousand feral buffaloes on the coastal plains of northern Australia, for example, has become a very emotional issue among Australian environmentalists, some of whom foretell the complete destruction of the environment if the uncontrolled feral herds are not destroyed.*

Soil Compaction

Water buffaloes have larger hooves than cattle of comparable size and thus they compact the soil less. But buffaloes often live in damp, boggy areas where their feet may compact soft soils. Also, buffaloes are creatures of habit and, when able, they set up fixed points for drinking, feeding, defecating, wallowing, and sleeping. Between the points they wear sharply defined trails in the vegetation and soil.

Wallowing

Possibly the water buffalo's greatest environmental limitation is its propensity to build wallows. In hot climates every buffalo will wallow at some time during the heat of the day if water is available. When they can, buffaloes will make their own wallows, enlarging a mud puddle by rolling in it or even using their heads to flip water out of a drinking trough and muddying the ground nearby.

---

*It is not at all clear, however, that the buffaloes (which have existed in the area for 150 years) are causing the observed environmental degradation. Other possibilities include: fire, climatic stress, overgrazing, and a variety of farming, hunting, and other human activities, especially the use of four-wheel-drive vehicles. Thousands of wild pigs also share the area, along with crayfish that burrow into and weaken the levees that keep out the sea, something for which the buffaloes have been blamed.

Kandep, Papua New Guinea. In creating a wallow, buffaloes may contribute to land destruction. (J. Schottler) . . . .

. . . However, water buffaloes can also benefit the environment. Shown here in Israel they are being managed to keep down swamp plants. (Gideon Shapira)

The pasture in the immediate area of the wallow is usually damaged by trampling and waterholes may become fouled, but buffaloes return to the same wallow day after day and do not build new ones indiscriminately. Thus, the muddied area is not a large proportion of the location in which they graze unless a large number of animals are confined in a small space.* In addition, man-made wallows can be dug at safe sites and the animals will use them. The problem of wallowing is therefore not generally a serious one.

Damage to Waterways

Because buffaloes often live near and enter water freely, they may cause erosion in ditches, river banks, canals, and levees. Also, their wallowing muddies the water, which may adversely affect some fish species and reduce the growth of algae. Buffaloes commonly urinate and defecate in the water, possibly creating a pollution hazard, although in most situations this contamination is likely to be minor.

The presence of this herbivore in natural waterways may reduce the number of water plants. Some plants are trampled, some eaten, and some underwater species are suppressed because the muddied water transmits less light. This (and several other of the buffalo's environmental effects) can be turned to advantage (see picture page 85) when, as often occurs, aquatic plants grow out of control and become obnoxious weeds.

Damage to Pastures

Water buffaloes have very strong jaws, and when forage is sparse they graze it close to the ground; this overgrazing can destroy a pasture. In addition, they eat virtually all available plant material (including many species that cattle shun), so that a densely stocked pasture can become completely defoliated. In northern Australia it has been found that, with time, buffaloes become accustomed to a given pasture, and unless fences are strong they will instinctively return to it until the forage has been depleted.

The buffalo's inclination to eat many plants can be used to improve the environment and suppress growth of coarse weedy species of plants. On the Sepik Plains in Papua New Guinea buffaloes are being used to graze and suppress sedges (*Cyperus* species); as a result, the more desirable *Paspalum* species are beginning to appear. At Mount Bundy in northern Australia native pastures are being improved on a commercial scale by overstocking them with

*At Gainesville, Florida (possibly because of its subtropical but not hot climate), a herd of 52 buffaloes concentrated in a one-hectare field did not attempt to build a wallow at all. (Information supplied by H. Popenoe.)

buffaloes. The animals reduce or completely eliminate spear grass and other weeds—even those with thorns—and thus foster the survival and growth of introduced forage legumes such as stylo (*Stylosanthes guianensis*).* In Sri Lanka buffaloes have been used to graze out the vigorous tropical grass *Imperata cylindrica*.†

Damage to Trees

Buffaloes instinctively rub against trees (and walls and fences), eagerly browse leaves, and sometimes nibble bark, so they damage trees more so than cattle. In northern Australia it has been noted that each "family herd" of feral buffaloes selects one or two trees for rubbing against so that the rubbing damage is confined to them.‡

Selected Readings

Rajapske, G., 1950. Death of illuk. *Ceylon Coconut Quarterly*, 1:7–9.

Tulloch, D. G. 1969. Home range in feral water buffalo. *Australian Journal of Zoology* 17:143–152.

Tulloch, D. G. 1970. Seasonal movement and distribution of the sexes in the water buffaloes in the Northern Territory. *Australian Journal of Zoology* 18:399–414.

Tulloch, D. G. 1975. Buffalo in the northern swamplands. *Proceedings III of the World Conference on Animal Production*. Theme 1, Paper 7.

Tulloch, D. G. 1977. Some aspects of the ecology of the water buffalo in the Northern Territory. In: *The Australian buffalo—a collection of papers*, edited by B. D. Ford and D. G. Tulloch. *Technical Bulletin* No. 18, Department of the Northern Territory, Animal Industry and Agriculture Branch, Australian Government Printing Service, Canberra, Australia.

*Tulloch, 1977.
†Rajapskse, 1950.
‡Tulloch, 1969.

11 Recommendations and Research Needs

This report has outlined the water buffalo's apparent merits, but most of the statements made about the animal are based on empirical observations. Many of its most exciting and potentially valuable features have not been subjected to the careful scrutiny needed to confirm their validity.

Despite the fact that there are 130 million or more water buffalo in the world, research on the animal is scanty and limited to only a few situations and sites. Quantitative information (especially for the various breeds), tests, trials, and comparison studies are needed.

The research to be done on the water buffalo offers scientific challenges that can be undertaken in laboratories and experiment stations in most parts of the world and in many disciplines: breeding, physiology, microbiology, veterinary science, nutrition, food science, dairy science, and other fields. Water buffalo research is an area worthy of financial support by philanthropic institutions and international development agencies concerned with problems of food and resource shortages. The dominant role of the buffalo in the rural economies of Egypt and Asian countries offers the opportunity for buffalo research that can bring improvements quickly and easily to the rural poor. For other countries the water buffalo is an untapped resource, and they should test its productivity on native pasturelands, marshy lowlands, hot and humid areas where cattle do not thrive, and on areas prone to cattle diseases and parasites that are difficult to control.

Specific recommendations follow.

Comparison with Cattle

Animal scientists worldwide should undertake trials to compare growth rate, feeding, nutrition, breeding, and other aspects of buffalo and cattle performance.

Cattle and water buffaloes are obviously different animals. Each has its own limitations and advantages, and each deserves to be studied in its own right. Perhaps the quickest way for animal scientists to experience for themselves the merits of the water buffalo is to conduct their own comparative trials with buffaloes and cattle in their areas. The results will provide local guidance and will help extend recognition of the buffalo's value, especially under difficult conditions where it may exceed cattle in productivity and profitability.

Germ Plasm Preservation

Urgent action is needed, especially in Southeast Asia, to preserve and protect outstanding buffalo specimens.

In some countries (Thailand, Malaysia, and Indonesia, for example) buffalo populations are decreasing dramatically. High demands for meat are causing slaughter at younger and younger ages. Much of the meat is exported to restaurants and markets in Singapore and Hong Kong. Unfortunately, the largest and quickest growing animals are often selected for slaughter. This results in the loss of a major genetic resource, which is compounded by the practice of castrating the largest males to make them more tractable as work animals. Ten years ago in Thailand it was common to find buffalo weighing 1,000 kg; now it is hard to find 750-kg specimens. A similar situation exists in the Philippines where there is no dearth of good breeding stock, but butchers are paying such high prices that farmers are selling even quality animals for slaughter. In northern Australia, where some of the bulls weigh almost 1,200 kg, the largest animals are being shot for meat, hides, pet food, or sportsmen's trophies.

A large number of high-yielding buffaloes are taken each year to big cities in India (for example, Bombay, Calcutta, Madras) for milking. At the end of lactation many are returned to the villages, but many others are slaughtered, rather than being fed and rebred. This creates a huge loss of valuable germ plasm. In many locations most of the largest animals have already been lost. Only urgent action will protect those remaining.

International Shipment

Buffalo quarantine stations should be organized in "disease-free" areas to develop buffalo germ plasm pools for international exchange.

The importation of buffaloes presents difficulties for any government, researcher, or farmer wishing to obtain the animals for the first time or for breeding purposes. Quarantine laws make it extremely difficult and expensive to exchange genetic resources.

Australia is one of the few nations where there are large numbers of water buffalo in an area free of the major animal diseases. Papua New Guinea,

Nigeria, Colombia, Venezuela, and other nations have taken advantage of this and have imported Australian buffaloes. But Australian herds are all Swamp buffaloes* and so breeding centers should be set up also (in Sri Lanka and Italy perhaps) where importers can obtain River-type (including Mediterranean) buffaloes.

Genetic Improvement

Worldwide efforts should be made to select superior buffalo bulls and cows for breeding.

Performance testing, leading to the mass selection of superior animals, deserves high priority. Virtually all buffalo breeding is haphazard and unplanned. Village animals graze together and matings are usually not controlled, observed, or recorded. Thus, the full genetic potential of the water buffalo is not being realized.

A massive selection program is needed to bring about genetic progress. For each breed, bulls and cows with the potential for improving production of meat and milk and increasing draft power should be identified and used for breed improvement. However, the wide variations between the characteristics of individual animals may make exceptional genetic advances difficult to achieve quickly.

Important traits for culling and selection include behavior, temperament, reproduction rate, easy milk letdown, average daily gain in weight or weight at a given age, carcass quality (for example, large hindquarters), and milk production, as well as strength and endurance for work.

Crossbreeding of Swamp and River buffaloes is a potentially important route to genetic improvement. The progeny reportedly show hybrid vigor (heterosis) in milking ability, fertility, meat production, and working ability. Infusing genes for high milk production into the Swamp buffalo, now used mainly for meat and work, creates the potential for a triple-purpose animal.

The use of artificial insemination and deep-frozen semen should be a major help in upgrading the buffalo. Moreover, the transport of live embryos (rather than neonatal animals) for implantation in the uterus of surrogate mothers could be important for water buffalo. It seems unlikely, however, that buffalo embryos can be implanted in cattle.

Most genetic selections should be made in Asia where 97 percent of the world's water buffaloes are located. The improvements will depend on how accurately bulls can be identified, selected, and mated. Performance and

---

*The island of Guam is also a safe source of Swamp animals, although the feral herds there are depleted; only 300 or 400 animals were left on Guam in 1978.

progeny testing is sorely needed at research stations as well as "on the farm." Governments should also institute bull-loan or artificial-insemination programs as a means for upgrading the village herds.

Comparison of Breeds

The relative merits of the various buffalo breeds should be determined.

Little or no information is available on the comparative performance of the different buffalo breeds in various environments, especially the 17 or so River breeds in the Subcontinent and the Egyptian and Mediterranean breeds. Comparison trials of the breeds and breed-crosses are needed in a wide range of climates from the humid tropical to the temperate. In addition, the cytogenetic, immunogenetic, and inheritance relationships of breeds should be clarified.

The panel encourages countries such as India and Pakistan that have a number of buffalo breeds (for example, Murrah, Surti, Jafarabadi, Mehsana, and Nili/Ravi) to set up experimental farms for scientific reproduction of superior specimens. Substantial research benefits as well as profitable economic returns from using and exporting some of them would be realized.

Meat and Milk Research

Research and demonstration is needed to foster the widespread consumption of buffalo meat and milk.

Buffalo milk, cheese, and other dairy products are considered outstanding foods in all locations where they are produced. Taste tests so far have indicated that buffalo meat is similar or slightly superior to beef produced under the same conditions.

A specific need is to feed the male calves and use them for meat. Many are now slaughtered at a young age and light weight. Research that provides either a partial or complete milk substitute for feeding calves would have a major impact on meat supplies and farmer income. Diets being developed for calves in Egypt incorporate such ingredients as whey, soybean meal, corn flour (corn starch), and yeast.*

Other research topics include:

• The effect of climate, thermoregulation, and wallowing on meat and milk production;

*Information supplied by M. El Ashry.

- The meat characteristics of each of the breeds and the differences be-
tween them;
- Milk production and quality for each of the breeds;
- Adapting buffaloes to machine milking by genetic selection or by de-
signing new milking machinery;
- Developing new or improved milk products (such as yogurt, cottage
cheese, and hard cheese); and
- Banking genetically superior germ plasm for later use.

Work Research

**The panel recommends research on new harnesses to replace the omni-
present yoke.**

As already noted, the wooden yoke, which has not changed in 1,500 years
or more, is an inefficient harness. Research is needed to adapt horse collars,
hames, breast straps, and other devices for the buffalo. Because much of the
farm power in Asia comes from buffaloes, the impact of improved harness
could be dramatic, widespread, and of enormous value to millions of small
farmers there. If the experiments in Thailand described earlier are an indica-
tion, the farm power in Asia could be increased by 25 percent overnight with
the adoption of an improved harness. The buffalo will continue to be the
small farmer's "tractor," so the benefits from improved harnesses are likely to
continue for a long time.

There are 13 million buffalo and bullock carts in India and 20 million In-
dians are engaged in the business of road haulage. Application of appropriate
technology would eliminate the archaic wooden wheels, axles, and heavy
carts and substitute lightweight carts, perhaps with such features as metal
wheels, pneumatic tires, ball bearings, and fixed axles. With such improvements, loads might be increased and hauled over longer distances at greater
speed and with less work.

Trials in New Areas

**Testing of water buffalo production is needed in many areas where the
animal is not known.**

A seemingly adaptable animal, the water buffalo should be productive
throughout the earth's warm temperate, subtropical, and tropical zones. Different breeds may adapt differently to extremes of heat, humidity, and cold,
and this needs further study.

The United States, the Mediterranean Basin of Europe, and some of the
more temperate European areas like southern England are worth considering

for water buffalo trials. In the Southern Hemisphere River buffaloes are already found as far south as 25° latitude in Brazil's São Paulo State (where large herds are raised); an experimental herd of Swamp buffaloes has performed outstandingly in Brisbane, Australia (27°S); and there are a few Swamp buffalo in South Australia and Victoria (35°S or more). There is good reason to believe that water buffaloes may be productive in all of the states in Australia, New Zealand's North Island, South Africa, Argentina, and other warm temperate areas of the Southern Hemisphere.

The biggest void in the water buffalo map is virtually the entire continent of Africa. It seems a paradox that the buffalo—Egypt's most important domestic animal—is not farmed commercially in any other African country. Experimental herds have been introduced to Nigeria, Uganda, Mozambique, Tanzania, and other countries in the past, and the initial success of three of them is described in Appendix A.

The water buffalo, with its tolerance for heat, disease, poor-quality feed, and mismanagement, appears to have outstanding promise for African nations such as Sudan, Tunisia, Morocco, Senegal, and The Gambia as well as all nations south of the Sahara (Nambia perhaps being an exception).

More specific aspects of environmental tolerance deserving research attention are:

• The physiology of the buffalo's response to heat, cold, humidity, and other environmental factors;
• The effect of climate on growth, reproduction, milk production, health, respiration rate, behavior, and carcass quality; and
• Measurement of the calorific efficiency and chemical composition of all breeds of buffaloes and cattle to determine the environments best suited to each breed.

Nutrition Research

Despite observations of the buffalo's ability to utilize poor quality forage, research is needed to learn how the animal does it.

This research should:

• Establish the buffalo's nutrient requirements by breed, sex, age, and weight for maintenance, growth, reproduction, lactation, and work;
• Determine voluntary forage consumption and the nutrient utilization of different forages in various stages of maturity;
• Examine rumen microbiology and fermentation, the rate of digestion, production, and absorption of volatile fatty acids produced in the rumen, and utilization of energy, nitrogen, vitamins, and minerals;

- Develop milk replacements for early-weaned calves;
- Observe the current village-level feeding of low-quality forages to learn their nutrient requirements, nutrient deficiency diseases, and nutrient supplementation needed;
- Study the utilization of concentrated, high-energy feeds (especially by-product feeds) to determine the upper limits of buffalo growth and productivity (milk, meat, and work) and carcass quality;
- Compare various breeds of water buffaloes and other ruminants to determine possible differences in nutritional requirements and performance; and
- Apply economic research to production practices, including night feeding of cut forage and the use of improved pastures.

Management

Research to improve management practices could benefit small farmers, ranchers, and feedlot feeders alike.

Little is known about the farm management factors that influence water buffalo productivity. Studies and extension literature for farmers are needed on subjects such as:

- Proper fencing;
- Procedures for dehorning and for preventing horn growth in calves;
- Grazing management methods;
- Preservation of forages;
- Methods and effect of castration;
- Methods and effect of spaying;
- Breeding methods;
- Milking methods;
- Artificial insemination, methodology and use;
- Methods of handling animals under grazing conditions or in feedlots;
- Age for weaning under different conditions;
- Mineral and protein supplementation methods;
- Semen freezing;
- Ova transplantation methods;
- Cryobranding and other methods of animal identification;
- Control of internal parasites; and
- Disease prevention and treatments.

Health

Research is needed in areas where uncertainties about buffalo diseases are hindering the animals' use.

The research should:

• Develop control and management practices to prevent specific infections and parasites for buffalo groups such as calves, yearlings, heifers, pregnant cows, newly calved cows, and bulls;
• Elucidate the factors (genetic, nutritional, management, disease) causing losses of newly born calves;
• Prevent and control the major diseases and parasites of the buffalo: hemorrhagic septicemia, brucellosis, tuberculosis, foot-and-mouth disease, sarcoptic mange, fascioliasis, nematode parasites, rinderpest and "rinderpest-like" diseases, nephritis, and conjunctivitis;
• Develop herd-health programs and specific disease-control programs for the various production systems;
• Determine the level of susceptibility to trypanosomiasis, a disease that might mitigate against introducing the animal widely to Africa; and
• Define the buffalo's role in malarial ecology.

Reproduction

Research is needed to improve water buffalo reproduction.

Research topics should include:

• Physiology and deep-freezing of buffalo semen. (Although it is possible to freeze the buffalo semen now, further improvements are needed to achieve higher conception rates.)
• Incidence of sub-estrus and anestrus. This is fairly high in buffaloes. Work is needed to determine the factors contributing to this problem and find solutions applicable in the field. A simple, inexpensive test for the routine diagnosis of estrus is needed. It could ensure that buffaloes are inseminated at the optimal time and could lead to the possible synchronizing of estrus in groups of animals as well as the elimination of seasonal breeding.
• Seasonality of breeding. Investigations are needed into seasonal effects on the intensity and deviation of estrus in buffaloes. Most of the buffaloes in northern India and Pakistan, for example, calve between July and December, causing scarcity of milk in the summer season and a flush of production in the winter months (this phenomenon causes serious marketing problems).

- Low libido and low semen yield in buffalo bulls. Semen yield in buffalo bulls is less than half of the yield from cattle bulls.
- Effect of season and other factors on semen quality.
- Variation in the freezability of semen from different bulls.

Dissemination of Information

The panel recommends that two water buffalo publications be produced.

These should be:

- An international water buffalo newsletter. It is important to maintain communication among researchers working with the water buffalo in far-flung research stations, universities, missions, and villages. Research findings may not be widely shared if technical animal science journals and the one or two national newsletters now available remain the only source of water buffalo information. A newsletter would bring together results from different parts of the world. It would provide rapid exchange of information as well as a forum for informal opinions, observations, and preliminary experimental data that are usually not accepted by journals.
- A formal journal of water buffalo research.

In addition, other methods for disseminating water buffalo information are to be encouraged.

Appendix A

Water Buffalo in Africa

For the water buffalo, Africa is the unknown continent. Apart from Egypt, where the "gamoosa" is a major livestock resource, there have been few recorded experiences with the animal in Africa. While this may be attributed to disease, chances are it is because of historical oversight.

Small herds of buffaloes were recently introduced to Uganda, Tanzania, and Nigeria, but because there were so few animals no firm conclusions can be drawn. Initial observations, however, suggest that the water buffalo could have an important future role in Africa, south of the Sahara. There seems little reason to believe that they won't thrive there as they have done elsewhere in the tropics.

The following statements were provided by researchers involved in the enterprises.

Uganda

"Uganda imported a herd of 12 buffalo cows and one bull in 1969. Grazed on poor-quality unimproved pasture they received neither supplementary feeding nor any preferential treatment over the East African zebus kept with them. The herd stayed 7 years at Entebbe before being moved to another part of the country. They proved to be very efficient converters of the low-protein, high-fiber fodder. During this time the herd grew to 40 adult cows and 2 bulls (all other steers were slaughtered). The cows calved every year. The calves matured at 2 to 2½ years; the zebu calves took at least 3 years to mature. Buffalo calves were much heavier than the zebus of comparable age.

"The buffalo cows averaged 7 liters of milk per milking. In taste it was preferred to zebu milk. Also, the quality of the meat after slaughter was much better than that of the zebus.

"A major advantage the buffaloes had over the cattle was that they were remarkably unaffected by diseases endemic to the area. Apart from a few calves that died of diseases that also killed cattle on the farm at Entebbe, the buffaloes resisted tick-borne diseases, the biggest killers of cattle in Uganda, and were unaffected by the virulent strains of East Coast Fever (Theileriosis), Uganda's most serious endemic animal disease."*

*Information supplied by G. L. Corry, Director, Veterinary Research Services, Entebbe, Uganda.

Nigeria

Early in 1976, 194 buffaloes aged 6-9 months were imported from the environs of Darwin in Australia's Northern Territory. The animals settled down quite well, and by the end of 1978 the stock population had risen to about 320, the adults weighing about 400 kg with about 80 percent fertility. The animals were in excellent health when an outbreak of streptothricosis occurred early in 1979. Because of the intractable nature of this rare disease of cattle, sheep, and pigs, the apparent spread of the disease in the confined area of the ranch, and the persistence of the organism, the animals were slaughtered.*

Tanzania

Tanzania introduced 21 buffalo heifers and 2 buffalo bulls from Egypt in 1968 and 1970. By 1977 the herd had grown to 150 animals. They were kept at the Livestock Production Research Institute at Mpwapwa, a site representative of central Tanzania.†

Over a period of 10 years the buffalo were studied and compared with crossbreeds of local Mpwapwa cows and Friesian steers. On average, the buffaloes conceived and calved 4 months earlier than the cattle, their mean calving intervals were about a month shorter, their birth and weaning weights were 1.5 times higher, and their daily weight gain was about double that of the crossbred cattle. The milk yield from the buffaloes was only about two-thirds that of the crossbred cattle, but their milk's high butterfat content meant that the overall butterfat production was higher in buffaloes.

Figures reported are shown below:

	Buffalo	Crossbred Cattle
Mean ages at first calving (months)	33	37
Mean birth weight (kg)	39	27
Mean daily weight gains (kg)	0.35	0.25
Mean weaning weight (kg)	88	60
Calving interval (days)	359	398
First lactation milk production (kg)	1,000	1,539
Peak milk yields (kg)	1,775	2,000
Butterfat content of milk (percent)	8.6	4.8

Source: Katyega, P. M. J. 1981. Production traits in Egyptian water buffaloes and Mpwapwa cattle. Tanzania Veterinary Bulletin 3:26-34 and Katyega, P. M. J., Masoud, A. J., and Kobo, E. 1980. Body and carcass characteristics of Egyptian water buffalo and Mpwapwa X Friesian steers raised in Central Tanzania. Tanzania Veterinary Bulletin 2:86-90.

*Information supplied by Dr. J. E. Erhiaganoma, Chief Veterinary Officer, Ministry of Agriculture and Natural Resources, Benin City, Bendel State of Nigeria.
†Mpwapwa is located at an elevation of 1,000 m. It receives an annual rainfall of 700 mm during a single season between late November and early May. Mean minimum temperatures of 13°C occur in June, and mean maximum temperatures of 26°C occur in November.

In a second study, 30 buffaloes and 34 Mpwapwa cattle* (not the cross-breeds) were compared.† The calving intervals were 396 ± 126 days for buffaloes and 437 ± 109 days for cattle. The buffaloes first conceived at mean ages of 23-25 months (matings occurred in both wet and dry seasons), the cattle at 27-36 months. The buffaloes calved at 35 months, the cattle at 46 months.

On farms at Ruvu and Mabuki the calving rate was 63 percent, and the mortality rate 1.2-6.9 percent.‡ The buffalo's overall milk yields averaged 1,237 kg in 225 days, or 5.5 kg per day.

The results of the introduction of buffalo to Tanzania are considered "quite encouraging and the future of buffaloes in Tanzania will be bright." § Future plans call for increasing the herd to about 200 animals with a view toward establishing another buffalo herd. In addition, 5 buffalo sires have been selected in Egypt for shipment to Tanzania.

---

*Mpwapwa cattle are a Tanzanian breed with 55 percent Asian, 35 percent African, and 10 percent European blood.
†Shoo, R. A. 1980. A study of the performance of Egyptian water buffaloes. Student project, Animal Production Department, Faculty of Agriculture, Forestry and Veterinary Medicine, University of Dar-Es-Salaam, Morogoro, Tanzania.
‡Rakha, A. 1980. Water Buffalo Production, Tanzania. Report of the Technical Cooperation Program TCP/URT/9002, Food and Agriculture Organization of United Nations, Rome, Italy.
§Oloufa, M. M. 1981. The future of water buffaloes in Tanzania. Paper presented at the Tanzania Society of Animal Production Eighth Scientific Conference, Arusha, Tanzania, May 26-29, 1981.

Appendix B

Research Contacts

A directory of names in buffalo studies and research is available from the International Buffalo Information Center, Kasetsart University Library, Bangkhen, Bangkok 10900, Thailand.

Australia
Animal Industry Branch, Department of Primary Production, P.O. Box 5160, Darwin, Northern Territory 5790 (D.R. Thomson, B. Ford)
Berrimah Agricultural Research Station, Darwin, Northern Territory
Coastal Plains Research Station, near Darwin, Northern Territory
D. G. Tulloch, P.O. Box 38841, Winnellie 5789, Northern Territory
University of Queensland, Department of Animal Husbandry, St. Lucia, Queensland (D.D. Charles)

Brazil
Associacáo de Criadores de Búfalos do Brasil, Cx Postal 832, São Paulo (President, Dr. Paulo Joaquim Monteiro da Silva)
Centro de Pesquisa Agropecuária do Trópico Umido (Agricultural Reseach Center of the Humid Tropics), Caixa Postal, 48, Belém, Pará, 66000
Faculdade de Medicina Veterinária e Zootecnia (Faculty of Veternary Medicine and Zootechnology), Universidade Estadual "Júlio de Mesquita Filho," Campus de Botucatu, Estado de São Paulo
Instituto de Zootecnia (Institute of Zootechnology), Nova Odessa, São Paulo (Prof. Alberto Alves Santiago)
Dr. Geraldo Mosse, Rua Apinages 716-CEP, 05017, São Paulo, S.P.
Dr. Walmore Muller Lacort, Ministerio de agricultura Explanada des Ministerios, Bloco 8-50 Andar Sala 518, 70.000 Brasilia, D.F.

Bulgaria
Academy of Agricultural Sciences, Sofia (P. Ivanov, Z. Zahariev)
Buffalo Basis, Livestock Breeding Research Institute, Shumen (Dr. D. St. Polikhronov, Director; Dr. A. Alexis)

China
Department of Animal Husbandry and Veterinary Medicine, Nanking Agricultural College, Nanking (Professor Sieh Chen-Hsia)
Kwangsi Agriculture College, Nanning, Kwangsi (Professor Wang Pei-Chien)
Livestock Farm, Lu Hsu, Pinyang County, Kwasngsi Chang Autonomous Region
Research Institute for Animal Science, Nanning, Kwangsi (Dr. Chou Chi-Sheng)
Singchow Dairy Farm, Kwangtung Province

Costa Rica
José Luis Pacheco, Director, Dept. Formento, 8502 JAPDEVA, Puerto Limon

Egypt
Ain Shams University, Kasr-El-Zaafran, Abbasiyah, Cairo (M.A. El-Ashry, A.M. El-Serafy)
Department of Animal Production, College of Agriculture, Cairo University, Cairo (Dr. M.T. Ragab)
University of Alexandria, 22 Al-Gueish Avenue, Shatby, Alexandria (K. El-Shazly)

Federal Republic of Germany
Dr. R. Dunkel, Herzeleidstrasse 39, 5330 Königswinter 41, Bonn
Prof. Dr. H. Fischer, Insitut fur Tropische Veterinarmedizin, Wilhelmstr. 15, 6300 Giessen
Gesellschaft fur Agrarentwicklung (GAE), Meckenheimer Allee 113, 5300 Bonn 1

Hungry
J. Kovacs, Department of Animal Husbandry, Agricultural University, Deak F. -u16, 8361-Keszthely

India
Andhra Pradesh Agricultural University, Hyderabad, A.P. (Director of Research)
Animal Husbandry Commissioner, Government of India, Ministry of Agriculture and Irrigation, Krishi Bhavan, New Delhi
Animal Sciences, Director General, Indian Council of Agricultural Research, Krishi Bhavan, New Delhi
Central Buffalo Breeding Farm, Sambalpua, PO: Sunbeda, Distt. Kerapur, Orissa
Central Frozen Semen Bank, Hessarghatta, Bangalore, Karnataka
Central Murrah Breeding Farm, Almadi, Tamil Nadu
Central Surti Buffalo Farm, Dhamrod, Gujarat
Chief Superintendent, Government Livestock Farm, Hissar, Haryana
College of Veterinary and Animal Science, Chandra Sekhar Azad University of Agriculture and Technology, Mathura Campus, Mathura, U.P.
Co-ordinator, All India Buffalo Improvement Project, Natioal Dairy Research Institute, Karnal, Haryana, (Dr. V. N. Tripathi)
Director, Military Dairy Farm, Army Headquarters, Quatermaster, General Branch, DHQ PO New Delhi 11
G.B. Pant, University of Agriculture and Technology, Pantnagar, U.P.
Government Livestock Farm, Anjora Distt: Durg, Madhya Pradesh
Gujarat Agricultural University, Reproductive Biology Research Unit, PO Anand Agricultural Institute, Anand, Gujarat
Haryana Agricultural University, Hissar, Haryana
Indian Veterinary Research Institute, Izatnagar, Barcilly, V.P. (Dr. S.K. Ranjan)
Jawaharlal Nehru Krishi Vishwa Vidyalaya, Jabalpur, 482004, M.P.
Kaira District Cooperative Milk Producers Union Ltd., Anand, Gujarat (Dr. V. Kurien, Chairman)
National Dairy Development Board, Anand, Gujarat
National Dairy Research Institute (ICAR), Karnal 132002, Haryana (S.P. Arora, Dr. D. Sundaresan, Director)
Punjab Agricultural University, Ludhiana, Punjab (Dr. J.S. Ichhponani, Dr. S.K. Misra)
University of Agricultural Sciences, Dharwar Campus, Dharwar, Karnataka
University of Udaipur, Udaipur, Rajasthan

Indonesia
Balai Inseminasi Buatan Lembang, Jawa Barat (Dr. R.D. Simangunsong)
Directorate of Animal Husbandry, Department of Agriculture, JI Salemba Raya 16, Jakarta (Dr. Jaman Zailani)
Pusat Penelitian dan Pengambangan Ternak (Centre for Animal Rsearch and Development), Bogor

Iraq
Animal Rearing Station, Abu Ghraib, Baghdad

Italy
Animal Production and Health Division, FAO,00100 Rome
Associazione Provinciale Allevatori, via Redentare, Caserta (Professore de Franciscis, Presidente)
Instituto de Ricerche soll'Adattamento Bovini e dei Bufali, Ponticelli, Napoli (Director: Prof. Lino Ferrara)
Instituto Sperimentale per la Zootecnica, via Onofrio Panvino, Rome (Prof. A. Pilla)
Instituto Sperimentale per la Zootecnice di Roma, via Salaria 31, 00016, Monterotondo Scalo (Dr. Augusto Romita)
Ugo Jemma, Torre Lupara, 81050 Pastorano, near Caserta

Japan
H. Shimuzu, The University of Tsukuba, Sakura-mura niihar-gun ibaraki-ken 300-311

Malaysia
Central Animal Husbandry Station, Kluang
Faculty of Veterinary Medicine and Animal Science, University Pertanian, Serdang, Selangor
Malaysian Agricultural Research and Development Institute, Serdang, Selangor

Nepal
Artificial Insemination Project, Tripureshwar, Kathmandu (Dr. A.C. Gupta)
Livestock Division, Khumaltar, Lalitpur, Kathmandu (Dr. N.D. Joshi)

Pakistan
College of Animal Husbandry, Lahore (Dr. Mohammad Shafi Chaudhry)
Department of Animal Reproduction, University of Agriculture, Faisalabad (Dr. Rashid Ahmad
    Chaudhary)
Livestock Production Research Institute, Bahadurnagar District, Sahiwal, Punjab (Dr. S.K.
    Shah, Director)
Punjab Veterinary Research Institute, Lahore 13

Panama
Hugo Giraoud, Apartado 60-2745, El Dorado

Papua New Guinea
Department of Primary Industry, ERAP, P.O. Box. 348, Lae (John H. Schottler)
Sepik Plains Livestock Station, Urimo via Wewak, E.S.P.

Philippines
Animal Science Department, Central Luzon State University, Muñoz, Nueva Ecija
College of Veterinary Medicine, Univeristy of the Philippines Systems, Diliman, Quezon City
Dairy Development Division, Bureau of Animal Industry, Sta. Mesa, Metro Manila (Dr. Conrado
    A. Valdez)
Dairy Training and Research Institute, University of the Philippines at Los Baños, College,
    Laguna 3720
Department of Animal Science, University of the Philippines at Los Baños, College, Laguna 3720
    (Dr. V. C. Momongon)
Documentation Center on Water Buffalo, University of the Philippines at Los Baños Library,
    College, Laguna 3720
Philippine Council for Agriculture and Resources Research, Livestock Research Division, Los
    Baños, Laguna 3732 (Dr. Alfonso N. Eusebio)

Portugal
W. Ross Cockrill, 591 Vale do Lobo, Almansil, Algrave 8100

Sri Lanka
Department of Animal Production and Health, Peradeniya (Dr. K. Balachandran)
School of Veterinary Science, University of Sri Lanka, Peradeniya Dr. B.M.A.O. Perrera)

Taiwan
Food and Fertilizer Technology Center, P.O. Box 22-149, Taipei City

Tanzania
Animal Production Department, Faculty of Agriculture, Forestry and Veterinary Medicine, University of Dar es Salaam, Morogoro (M.L. Kyomo)
DAFCO Farm, Ruvu
Director, Livestock Development Division, Ministry of Agriculture, Dar es Salaam
P.M.J. Katyega, Director, Livestock Production Research Institute, Private Bag, Mpwapwa M.M. Oloufa, FAO, Morogoro

Thailand
Kasetsart University, Department of Animal Science, Bangkok (Dr. Charan Chantalakhana)
Faculty of Veterinary Science, Chulalongkorn University, Henri Dunant Street, Bangkok 5 (Dr. Maneewan Kamonpatana)
Department of Obstetrics, Gynaecology and Reproduction, Faculty of Veterinary Science, Chulalongkorn University, Henri Dunant Road, Bangkok 10500
FAO Regional Office, Maliwan Mansion, Phra Atit Road, Bangkok 2 (Dr. B. K. Soni)
Office of Livestock Development Project, Tha Phra, Khon Kaen
National Buffalo Research and Training Center, Surin

Trinidad
Steve Bennett, 61 Mucurapo Road, Port-of-Spain

Uganda
G. L. Corry, Director, Veterinary Research Services, P.O. Box 24, Entebbe
S. Mugerwa, Faculty of Agriculture, MaKerere University, P.O. Box 7062, Kampala

United Kingdom
T. B. Begg, Resident Veterinary Surgeon, Howletts and Port Lympne Estates, Ltd., Port Lympne, Lympne, Kent CT21 4PD
P. N. Wilson, BOCM Silcock Ltd., Basing View, Basingstoke, Hampshire RG21 2EQ

United States
Wyland Cripe, Asst. Dean for Public Services, College of Veterinary Medicine, Box J-125, JHMHC, University of Florida, Gainesville, Florida 32610
J. Harrington, IRI Research Institute, Inc. One Rockefeller Plaza, New York, New York 10020
James Hentges, Department of Animal Science, Room 2104 McCarty Hall, University of Florida, Gainesville, Florida 32611
Jack Howarth, School of Veterinary Medicine, University of California, Davis, California 95616
Nels M. Konnerup, 609 East Iverson Road, Camano Island, Washington 98292
A. P. Leonards, P.O. Box 1094, Lake Charles, Louisiana 70602
John K. Loosli, Department of Animal Science, Room 2103 McCarty Hall, University of Florida, Gainesville, Florida 32611
Robert E. McDowell, Department of Animal Science, Frank B. Morrison Hall, Cornell University, Ithaca, New York 14853
Hugh Popenoe, International Programs in Agriculture, 3028 McCarty Hall, University of Florida, Gainesville, Florida 32611
William R. Pritchard, School of Veterinary Medicine, University of California, Davis, California 95616

Venezuela
Abelardo Ferrer D., Quinta Nueva Exparta, Avenida José Felix Rivas, San Bernardino, Caracas

Biographical Sketches of Panel Members

STEVE P. BENNETT received a Diploma in Agriculture, 1941, from the University of Guelph, Ontario, Canada, and Doctor of Veterinary Medicine Degree, 1948, from Colorado State University. He has been employed in mixed private practice in Trinidad for the past 30 years; this practice has always included the husbandry and welfare of water buffaloes.

CHARAN CHANTALAKHANA, Professor of Animal Science and Breeding, is at present seconded to the Ministry of Agriculture and Cooperatives as Secretary to the Deputy-Minister. He received his B.S. (1959), M.S. (1961), and Ph.D. (1968) from Iowa State University. He has worked with Kasetsart University, Bangkok, Thailand, since 1962. During 1963-67, he headed the university's Tankwang Experiment Station where he started his investigation on Thai indigenous cattle and crossbreeding. He has worked on problems concerning Thai Swamp buffalo production since 1971 and has been Project Coordinator and Chairman of Kasetsart University Buffalo Research Committee. He was Head, Department of Animal Science, Kasetsart University, from 1973-76, and Dean, Faculty of Natural Resources, Prince of Songkhla University, from 1978 to 1979.

DONALD D. CHARLES is Senior Lecturer in Meat Production at the University of Queensland. After completing his B.V.Sc. degree in 1953 he worked in the Meat Export Branch of the Commonwealth Department of Primary Industry. He joined the University in 1967, received an M.V.Sc. in 1971 for a system of grading beef carcass by specifications, and his Ph.D. in 1976 for a study of the age and breed effects on qualitative and quantitative characteristics of beef. He has studied carcass composition in the Swamp buffalo on different feeding regimes and has compared the changes in carcass composition with liveweight increase between the buffalo and cattle.

W. ROSS COCKRILL served with the Ministry of Agriculture in Britain and as a member of whaling expeditions to the Antarctic before joining the United Nations Food and Agriculture Organization (FAO) in 1953, from which he retired in 1975. He has traveled widely, lecturing and consulting, and has recently provided consultant services to the Asian Development Bank in Papua New Guinea, to the Swedish International Development

Agency in Vietnam, to FAO in India, Pakistan, and Thailand, and to numerous government and private organizations in the fields of animal production and health. He is the author of publications on whales and whaling, animal production, and disease control; editor of *The Husbandry and Health of the Domestic Buffalo* and *Animal Regulation Studies*; and author of *Antarctic Hazard, The Buffaloes of China*, and many papers on water buffaloes. He is a Fellow of the Royal College of Veterinary Surgeons in the U.K. and a Doctor of Veterinary Medicine of Zurich University.

WYLAND S. CRIPE, Associate Professor and Assistant Dean for Public Services at the University of Florida College of Veterinary Medicine, received an A.B. in food research at Stanford University in 1946 and a B.S. and D.V.M. at the University of California, Davis, in 1952. The following 16 years he was in private veterinary practice in the Sacramento Valley, California, area. During the 1970s he developed and implemented food-animal preventive medicine programs in Chile, Colombia, Dominican Republic, Ecuador, and Venezuela. He is the veterinary team leader of the water buffalo investigations at the University of Florida.

TONY J. CUNHA, Dean of the School of Agriculture, California Polytechnic University, Pomona, received his Ph.D. at the University of Wisconsin in 1944. He has lectured and studied livestock production in 37 foreign countries, has served on numerous committees, and has received many awards for research in animal nutrition. He served as Chairman of the Animal Nutrition Committee of the National Academy of Sciences (NAS). He has also served as a member of the NAS Board on Agriculture and Renewable Resources.

A. JOHN DE BOER, Agricultural Economist at Winrock International Livestock Research and Training Center, Morrilton, Arkansas, received a B.A. degree in 1966 from Colorado State University and his Ph.D. in agricultural economics from the University of Minnesota in 1972. His thesis research was carried out on cattle and water buffalo production under Thai village conditions. He was Lecturer in Agricultural Economics, University of Queensland, Brisbane, Australia, from 1972 to 1978; was a visiting scientist at the International Livestock Center for Africa in 1977; and has carried out a number of studies on livestock production problems in Asia. He has served as a consultant to the United Nations Food and Agriculture Organization and the Asian Productivity Organization. His major research interests focus on the application of economic analysis to problems facing small-scale livestock producers.

MAARTEN DROST, Chairman of the Department of Reproduction in the College of Veterinary Medicine, University of Florida, received his D.V.M. from Iowa State University in 1962. After two years each in private veterinary practice and the U.S. Army veterinary corps in San Jose, California, and Frankfurt, Germany, he joined the faculty of the University of California, Davis, where he specialized in cattle reproduction. His research has focused on the initiation of parturition and an embryo transfer in ruminants. He was a visiting professor at Cornell University in 1972-73 and at the State University at Utrecht, the Netherlands, in 1975-76. He has lectured in several European and South American countries and is a Diplomate in the American College of Theriogenologists.

MOHAMED ALI EL-ASHRY, Professor of Animal Nutrition at Ain Shams University, Shoubra El Kheim, Cairo, Egypt, received his Ph.D. from the Soviet Academy of Science in 1963 and his B.Sc. from Ain Shams University in 1958. He worked at the Higher Institute of Agriculture at Moushtohour during 1963-65, when he left to join the faculty of Ain Shams University. He received an award for distinguished research work at Ain Shams University in 1972. His research has focused mainly on ruminant nutrition, with emphasis on buffalo and sheep. He received a grant from the Egyptian Academy of Science and Technology for formulating and producing a milk replacer for buffalo calves, which has been used throughout Egypt since 1975.

ABELARDO FERRER D. is an agricultural engineer and veterinarian. He was a Professor of Zootechnology (1945-53), Director of the Animal Reproduction Center of Venezuela (1966-74), and Director of Livestock for the Ministry of Agriculture. Since 1974 he has been a partner and managing director of a farm in Apure, Venezuela, with 2,500 buffalo. Since 1967 he has been actively engaged in introducing buffalo to Venezuela.

JEAN K. GARNER retired as Chief of Logistics Division, USAID/Philippines, in 1970. He attended the University of Arkansas School of Engineering, was engaged in farming, highway construction and maintenance, and organized and served as president and manager of a farm equipment dealership. In 1951 he joined the United Nations Food and Agriculture Organization and was assigned to Pakistan as Farm Machinery Advisor to the Ministry of Agriculture. In 1955 he joined the U.S. Agency for International Development (USAID) and was assigned to Thailand as Agricultural Engineer Advisor to the Ministry of Agriculture. His research and development included work on improving the harness and implements for buffalo used on farms for traction power. He also served in similar positions in Vietnam, Indonesia, Thailand, and the Philippines.

JAMES F. HENTGES, JR., Professor of Animal Science in the College of Agriculture and Ruminant Animal Nutritionist in the Agricultural Experiment Station of the Institute of Food and Agricultural Sciences, University of Florida, Gainesville, received a B.S. in animal science at Oklahoma State University in 1948 and a Ph.D. in biochemistry-animal science at the University of Wisconsin in 1952. Currently he supervises beef cattle and water buffalo research at three research farms in Florida and teaches undergraduate, graduate, and continuing education courses. In 1975 he was chairman of the National Academy of Sciences panel that participated in workshops on aquatic weed problems in the Sudan and Egypt. His expertise lies in the fields of ruminant animal nutrition, physiology, and enterprise management.

JACK A. HOWARTH, Professor of Epidemiology and Preventive Medicine at the School of Veterinary Medicine, University of California, Davis, received a D.V.M. degree in 1944 from Colorado State University and his Ph.D. in comparative pathology from the University of California in 1950. He has been a University of California faculty member since 1950 and teaches the infectious diseases of farm animals to veterinary students. His research has focused on infectious diseases of animals, particularly those that are tick-borne. He has traveled widely in the tropics, lecturing and consulting on animal diseases and livestock production and has written numerous papers.

NELS KONNERUP was a Livestock Disease Specialist with the Agriculture and Fisheries Office of the U.S. Agency for International Development (USAID) before his retirement in 1979. He received a B.S. in 1941 and a D.V.M. in 1942 from Washington State University. Before joining the United Nations Food and Agricultural Organization (FAO) he was a veterinary practitioner for five years. After 10 years with FAO he worked for the U.S. Department of Agriculture as a veterinary analyst, for the Walter Reed Army Institute of Research, and joined USAID in 1966. He participated in numerous conferences on animal diseases in Asia, Africa, and Latin America and in 1964–65 he chaired the National Academy of Sciences-National Research Council study group on animal diseases in Africa. He has had long experience with the diseases of the water buffalo, notably in China.

ANTHONY P. LEONARDS received his B.S. in animal husbandry from the University of Southwestern Louisiana. At present he is President of Grex, Incorporated. Concurrent with his business activities, he became interested in water buffalo as a future prime meat source, and, while researching in Trinidad, Guam, Venezuela, and Australia, became the first, and currently the only, commercial breeder of water buffalo in the United States.

JOHN K. LOOSLI, Adjunct Professor at the University of Florida and Emeritus Professor of Animal Nutrition, Cornell University, received a B.S. from Utah State University in 1931, an M.S. from Colorado State University in 1932, and a Ph.D. from Cornell University in 1938. He served as Professor of Animal Nutrition at Cornell University from 1939 to 1974 and was Head of the Department of Animal Science from 1963 to 1971.

He has taught and conducted research at the University of Ibadan, Nigeria, and the University of Florida and was Fulbright Lecturer at Queensland University, Australia. His research has dealt with nutritional requirements, deficiencies, and efficiency of feed utilization of many species of farm and experimental animals, including water buffaloes. He has traveled extensively to study livestock nutrition and production in the United States, Latin America, Southeast Asia, India, and Africa.

JOSEPH C. MADAMBA is currently Director of the Philippines-based Asian Institute of Aquaculture. He earned his B.Sc. in agriculture at the University of the Philippines College of Agriculture, his M.Sc. at Cornell University, and his Ph.D. at the University of Illinois. An animal scientist by academic training, with specializations in animal nutrition and agricultural economics, his experience and areas of interest cover the area of livestock agri-business systems, agricultural development planning, research systems management, and aquaculture research and development. He has traveled extensively, studied a number of agricultural production systems, and has served as consultant to several agricultural development programs in various Asian countries. He has participated in numerous scientific, planning, and technical conferences concerning livestock, agricultural research management, agricultural education, transfer of technology, and farming systems development. He has been Associate Professor and Chairman of the Department of Animal Science, University of the Philippines, at Los Baños.

CRISTO NAZARÉ BARBOSA DO NASCIMENTO is Chief of the Agricultural Research Center for Humid Tropics (CPATU), Belém, Brazil, which belongs to the Brazilian Agriculture Research Corporation (EMBRAPA). He received a B.S. from the Amazon Agriculture School, Belém, in 1965 and an M.S. in animal science from Texas A & M University in 1968. From 1969 to 1976 he worked at the Northern Agriculture Research Institute (IPEAN). In 1976, with the transformation of IPEAN into CPATU, he continued his work as researcher at CPATU and became Chief in 1978. His research has been mainly on water buffalo feeding, management, and production systems.

DONALD L. PLUCKNETT, Professor of Agronomy and Soil Science at the College of Tropical Agriculture, University of Hawaii, received B.S. and M.S. degrees in agriculture and agronomy from the University of Nebraska in 1953 and 1957, respectively, and a Ph.D. in tropical soil science from the University of Hawaii in 1961. He has worked extensively in tropical crop and pasture research and has had broad international experience in tropical agriculture. He has been a consultant for many international groups, including work for the Ford Foundation on the Aswan Project in

Egypt, for the United Nations Food and Agriculture Organization, Consultative Group on International Agricultural Research, United States Agency for International Development (USAID), and the South Pacific Commission. From 1973 to 1976 he was Chief of the Soil and Water Management Division, Office of Agriculture, Technical Assistance Bureau, Agency for International Development, Washington, D.C. In 1976 he was awarded AID's Superior Honor Award for his activities in International Development. He has served on several National Academy of Sciences study panels.

HUGH L. POPENOE is Professor of Soils, Agronomy, Botany, and Geography and Director of the Center for Tropical Agriculture and International Programs (Agriculture) at the University of Florida. He received his B.S. from the University of California, Davis, in 1951 and his Ph.D. in soils from the University of Florida in 1960. His principal research interest has been in the area of tropical agriculture and land use. His early work in shifting cultivation is one of the few contributions to knowledge of this system. He has traveled and worked in most of the countries in the tropical areas of Latin America, Asia, and Africa. He is Chairman of the Board of Trustees of the Escuela Agricola Panamericana in Honduras, Visiting Lecturer on Tropical Public Health at the Harvard School of Public Health, and is a Fellow of the American Association for the Advancement of Science, the American Society of Agronomy, the American Geographical Society, and the International Soils Science Society. He is Chairman of the Advisory Committee for Technology Innovation and a member of the Board on Science and Technology for International Development.

WILLIAM ROY PRITCHARD, Dean, School of Veterinary Medicine, University of California, Davis, and Coordinator, International Agriculture Programs, University of California–System, received his D.V.M. from Kansas State University, Ph.D. from the University of Minnesota, and J.D. from the University of Indiana. He also has been awarded honorary D.Sc. degrees from Kansas State University and Purdue University. His major research interest has been the control of livestock diseases. He has been active in international development programs for over 20 years and has visited 65 countries as an adviser on education and research in agricultural areas.

J. THOMAS REID was Professor of Animal Nutrition at Cornell University where he also served as Head of the Department of Animal Science. He received a B.S. from the University of Maryland in 1941 and a Ph.D. in Nutrition and Biochemistry from Michigan State University in 1946. His major research and teaching interests were the comparative nutritional energetics and body composition of animals. He received the Nutrition

Award (1950) and the Borden Award (1957) of the American Dairy Science Association; the Merit Award (1965) of the American Grassland Council; and the Morrison Award (1967) of the American Society of Animal Science. In 1955–56 he received a Guggenheim Fellowship to conduct research at the University of Reading. He served on three task forces of the Council for Agricultural Science and Technology; two panels each of the Energy Research and Development Agency and the National Science Foundation; four committees of the National Academy of Sciences; and on the Federal Advisory Sub-Group to advise the President on Food and Nutrition in the United States. He worked in many countries, especially South America. Dr. Reid died on November 18, 1979.

JOHN H. SCHOTTLER earned a Bachelor of Agricultural Science degree at Melbourne University in 1967. He worked for a short period on pastures in the Northern Territory before moving to Papua New Guinea where he undertook research in cattle and buffalo. He then supervised the expansion of a buffalo ranch to receive buffaloes imported from Australia. He is currently Senior Animal Production Officer for ruminants in Papua New Guinea.

DALBIR SINGH DEV, Professor of Animal Science, Punjab Agricultural University, Ludhiana, Punjab, India, received a B.Sc. in agriculture and M.Sc. in plant genetics and biometry from Punjab University in 1955 and 1958 respectively; and M.S. and Ph.D. degrees in animal breeding from Ohio State University in 1961 and 1968 respectively. From 1968 to 1977 he was Professor of Animal Breeding and Head of the Animal Science Department of Punjab Agricultural University. In addition to teaching and extension education, this department deals with research on the genetics, nutrition, physiology, and management of farm livestock. The department maintains one of the best buffalo herds of Murrah and Nili/Ravi breeds. His research has been mainly in the area of animal genetics covering buffalo, cattle, and poultry.

ROBERT W. TOUCHBERRY, Dean, Department of Animal Science, University of Minnesota, St. Paul, received a B.S. in animal breeding in 1945 from Clemson College and an M.S. and Ph.D. in animal breeding and genetics from Iowa State College in 1947 and 1948 respectively. His interests include population genetics, quantitative genetics, and effects of crossbreeding on the growth and milk production of dairy cattle. He worked with advanced animal breeding in Sierra Leone and was a Fulbright Research Fellow in Denmark. An experienced animal breeder and geneticist, he was selected to ensure that the panel's conclusions were balanced. as he knows little about the water buffalo.

NOEL D. VIETMEYER, staff officer for this study, is Professional Associate of the Board on Science and Technology for International Development. A New Zealander with a Ph.D. in organic chemistry from the University of California, Berkeley, he now works on innovations in science that are important for developing countries.

Advisory Committee on Technology Innovation

HUGH POPENOE, Director, International Programs in Agriculture, University of Florida, Gainesville, Florida, *Chairman*

Members

WILLIAM BRADLEY, Consultant, New Hope, Pennsylvania

HAROLD DREGNE, Director, International Center for Arid and Semi-Arid Land Studies, Texas Tech University, Lubbock, Texas

ELMER L. GADEN, JR., Department of Chemical Engineering, University of Virginia, Charlottesville, Virginia

ANDREW HAY, President, Calvert-Peat, Inc., New York, New York (member through 1980)

CARL N. HODGES, Director, Environmental Research Laboratory, Tucson, Arizona

CYRUS MCKELL, Institute of Land Rehabilitation, Utah State University, Logan, Utah

FRANÇOIS MERGEN, Pinchot Professor of Forestry, School of Forestry and Environmental Studies, Yale University, New Haven, Connecticut

DONALD L. PLUCKNETT, Consultative Group on International Agricultural Research, Washington, D.C.

THEODORE SUDIA, Deputy Science Advisor to the Secretary of Interior, Department of Interior, Washington, D.C.

113

Board on Science and Technology for International Development
(JH-217D)
Office of International Affairs
National Research Council
2101 Constitution Avenue, Washington, D.C. 20418, USA

How to Order BOSTID Reports

Reports published by the Board on Science and Technology for International Development are sponsored in most instances by the U.S. Agency for International Development and are intended for free distribution primarily to readers in developing countries. A limited number of copies are available without charge to readers in the United States and other industrialized countries who are affiliated with governmental, educational, or research institutions, and who have professional interest in the subjects treated by the report. Requests should be made on the institution's stationary.

Single copies of published reports listed below are available free from BOSTID at the above address while the supplies last.

Energy

19. **Methane Generation from Human, Animal, and Agricultural Wastes.** 1977. 131 pp. Discusses means by which natural process of anaerobic fermentation can be controlled by man for his benefit and how the methane generated can be used as a fuel.

33. **Alcohol Fuels: Options for Developing Countries.** 1983. 128 pp. Examines the potential for the production and utilization of alcohol fuels in developing countries. Includes information on various tropical crops and their conversion to alcohols through both traditional and novel processes.

36. **Producer Gas: Another Fuel for Motor Transport.** 1983. 112 pp. During World War II Europe and Asia used wood, charcoal, and coal to fuel over a million gasoline and diesel vehicles. However, the technology has since been virtually forgotten. This report reviews producer gas and its modern potential.

39. **Proceedings, International Workshop on Energy Survey Methodologies for Developing Countries.** 1980. 220 pp. Report of a 1980 workshop organized to examine past and ongoing energy survey efforts in developing countries. Includes reports from rural, urban, industry, and transportation working groups, excerpts from 12 background papers, and a directory of energy surveys for developing countries.

Technology Options for Developing Countries

14. **More Water for Arid Lands: Promising Technologies and Research Opportunities.** 1974. 153 pp. Outlines little-known but promising technologies to supply and conserve water in arid areas. (French language edition is available from BOSTID.)

21. **Making Aquatic Weeds Useful: Some Perspectives for Developing Countries.** 1976. 175 pp. Describes ways to exploit aquatic weeds for grazing, and by harvesting and processing for use as compost, animal feed, pulp, paper, and fuel. Also describes utilization for sewage and industrial wastewater treatment. Examines certain plants with potential for aquaculture.

28. **Microbial Processes: Promising Technologies for Developing Countries.** 1979. 198 pp. Discusses the potential importance of microbiology in developing countries in food and feed, plant nutrition, pest control, fuel and energy, waste treatment and utilization, and health.

31. **Food, Fuel, and Fertilizer for Organic Wastes.** 1981. 150 pp. Examines some of the opportunities for the productive utilization of organic wastes and residues commonly found in the poorer rural areas of the world.

34. **Priorities in Biotechnology Research for International Development: Proceedings of a Workshop.** 1982. 261 pp. Report of a 1982 workshop organized to examine opportunities for biotechnology research in developing countries. Includes general background papers and specific recommendations in six areas: 1) vaccines, 2) animal production, 3) monoclonal antibodies, 4) energy, 5) biological nitrogen fixation, and 6) plant cell and tissue culture.

Plants

16. **Underexploited Tropical Plants with Promising Economic Value.** 1975. 187 pp. Describes 36 little-known tropical plants that, with research, could become important cash and food crops in the future. Includes cereals, roots and tubers, vegetables, fruits, oilseeds, forage plants, and others.

25. **Tropical Legumes: Resources for the Future.** 1979. 331 pp. Describes plants of the family Leguminosae, including root crops, pulses, fruits, forages, timber and wood products, ornamentals, and others.

37. **The Winged Bean: A High Protein Crop for the Tropics.** (Second Edition). 1981. 59 pp. An update of BOSTID's 1975 report of this neglected tropical legume. Describes current knowledge of winged bean and its promise.

47. **Amaranth: Modern Prospects for an Ancient Crop.** 1984. Before the time of Cortez grain amaranths were staple foods of the Aztec and Inca. Today this extremely nutritious food has a bright future. The report also discusses vegetable amaranths.

53. **Jojoba: New Crop for Arid Lands.** 1984. Describes *Simmondsia chinensis*, a North American desert shrub whose seeds are rich in a unique vegetable oil with considerable potential as an industrial raw material.

Innovations in Tropical Reforestation

27. **Firewood Crops: Shrub and Tree Species for Energy Production.** 1980. 237 pp. Examines the selection of species suitable for deliberate cultivation as firewood crops in developing countries.

35. **Sowing Forests from the Air.** 1981. 64 pp. Describes experiences with establishing forests by sowing tree seed from aircraft. Suggests testing and development of the techniques for possible use where forest destructions now outpaces reforestation.

40. **Firewood Crops: Shrub and Tree Species for Energy Production.** Volume II. 1983. A continuation of BOSTID report number 27. Describes 27 species of woody plants that seem suitable candidates for fuelwood plantations in developing countries.

41. **Mangium and Other Fast-Growing Acacias for the Humid Tropics.** 1983. 63 pp. Highlights ten acacias species that are native to the tropical rain forest of Australasia. That they could become valuable forestry resources elsewhere is suggested by the exceptional performance of *Acacia mangium* in Malaysia.

42. **Calliandra: A Versatile Small Tree for the Humid Tropics.** 1983. 56 pp. This Latin American shrub is being widely planted by villagers and government agencies in Indonesia to provide firewood, prevent erosion, yield honey, and feed livestock.

43. **Casuarinas: Nitrogen-Fixing Trees for Adverse Sites.** 1983. These robust nitrogen-fixing Australasian trees could become valuable resources for planting on harsh, eroding land to provide fuel and other products. Eighteen species for tropical lowlands and highlands, temperate zones, and semiarid regions are highlighted.

52. **Leucaena: Promising Forage and Tree Crop for the Tropics.** (Second Edition). 1984. 110 pp. Describes a multipurpose tree crop of potential value for much of the humid lowland tropics. Leucaena is one of the fastest growing and most useful trees for the tropics.

Managing Tropical Animal Resources

32. **The Water Buffalo: New Prospects for an Underutilized Animal.** 1981. 118 pp. The water buffalo is performing notably well in recent trials in such unexpected places as the United States, Australia, and Brazil. Report discusses the animal's promise, particularly emphasizing its potential for use outside Asia.

44. **Butterfly Farming in Papua New Guinea.** 1983. 36 pp. Indigenous butterflies are being reared in Papua New Guinea villages in a formal government program that both provides a cash income in remote rural areas and contributes to the conservation of wildlife and tropical forests.

45. **Crocodiles as a Resource for the Tropics.** 1983. 60 pp. In most parts of the tropics crocodilian populations are being decimated, but programs in Papua New Guinea and a few other countries demonstrate that, with care, the animals can be raised for profit while the wild populations are being protected.

46. **Little-Known Asian Animals with a Promising Economic Future.** 1983. 124 pp. Describes banteng, madura, mithan, yak, kouprey, babirusa, Javan warty pig, and other obscure, but possibly globally useful, wild and domesticated animals that are indigenous to Asia.

118

General

29. **Postharvest Food Losses in Developing Countries.** 1978. 202 pp. Assesses potential and limitations of food-loss reduction efforts; summarizes existing work and information about losses of major food crops and fish; discusses economic and social factors involved; identifies major areas of need; and suggests policy and program options for developing countries and technical assistance agencies.

30. **U.S. Science and Technology for Development: Contributions to the UN Conference.** 1978. 226 pp. Serves the U.S. Department of State as a major background document for the U.S. national paper, 1979 United Nations Conference on Science and Technology for Development.

For a complete list of publications, including those that are out of print and available only through NTIS, please write to BOSTID at the address above.

www.ingramcontent.com/pod-product-compliance
Lightning Source LLC
Chambersburg PA
CBHW021603210326
41599CB00010B/572